AF766365

Tensegrity

Tensegrity

Structural Systems for the Future

René Motro

KOGAN PAGE SCIENCE

London and Sterling, VA

To Madeleine
To Jean, Laure, Pascal, Sarah and... Theo

First published in Great Britain and the United States in 2003 by Kogan Page Science, an imprint of Kogan Page Limited

Apart from any fair dealing for the purposes of research or private study, or criticism or review, as permitted under the Copyright, Designs and Patents Act 1988, this publication may only be reproduced, stored or transmitted, in any form or by any means, with the prior permission in writing of the publishers, or in the case of reprographic reproduction in accordance with the terms and licences issued by the CLA. Enquiries concerning reproduction outside these terms should be sent to the publishers at the undermentioned addresses:

120 Pentonville Road
London N1 9JN
UK
www.koganpagescience.com

22883 Quicksilver Drive
Sterling VA 20166-2012
USA

© Hermes Science Publishing Limited, 2003
© Kogan Page Limited, 2003

The right of René Motro to be identified as the author of this work has been asserted by him in accordance with the Copyright, Designs and Patents Act 1988.

ISBN 1 9039 9637 6

British Library Cataloguing-in-Publication Data

A CIP record for this book is available from the British Library.

Library of Congress Cataloging-in-Publication Data

Motro, René, 1946-
Tensegrity : structural systems for the future / René Motro.
p. cm.
ISBN 1-903996-37-6
1. Geodesic domes. 2. Structural analysis (Engineering) I. Title.
TA660.D6M57 2003
624.1--dc21
2003007305

Typeset by Kogan Page
Printed and bound by Antony Rowe Ltd, Eastbourne
Transferred to digital printing 2005

Contents

Notations

Chapter 1

c	Length of tensioned elements (cables)
s	Length of compressed elements (struts)
r	Ratio s/c

Chapter 3

b	Number of elements
c	Length of tensioned elements (cables)
d_{ij}	Distance between nodes "i" and "j"
I	First end of element
j	Second end of element
l_{ij}	Fabrication length of element "i-j" to be inserted between nodes "i" and "j"
n	Number of nodes
r	Ratio s/c
s	Length of compressed elements (struts)
θ	Relative rotation of upper and lower polygonal faces
N_c	Internal force in cable
N_s	Internal force in strut

Chapter 4

n	Number of nodes
S	Number of struts
C	Number of cables
I	Irregular
R	Regular
SS	Spherical system
cc	Length of tensioned elements (cables)
ss	Length of compressed elements (struts)
H_c	Cable hyperboloïd
H_s	Strut hyperboloïd

Chapter 5

[A]	Equilibrium matrix
[C]	Connection matrix
$[C]^t_i$	i° column vector of transpose matrix $[C]^t$
$[D_x]$	Connectivity matrix
$[D^0_x]$	Connectivity matrix, for free nodes
$[D^0_{fx}]$	Connectivity matrix for fixed nodes
[Q]	Matrix of force density coefficients
$[Q^0]$	Matrix of pre-stress (self-stress) force density coefficients
$\{f_i\}$	Vector of external load applied on "i" along X-direction
{q}	Vector of force density coefficients
{T}	Vector of internal efforts
b	Number of members
c	Length of tensioned elements
c_{mk}	Coefficient of connection matrix
f_{ix}	Component of external load {fi} applied on "i" along X-direction
l^0_j	Reference length
m	Number of internal mechanisms
N	Number of degrees of freedom
n	Number of nodes
n_{lx}	Number of free nodes
q_j	Force density coefficient of element "i"
r	Ratio s/c ("single parameter")
r_A	Rank of equilibrium matrix
s	Length of compressed elements
ss	Number of self-stress-states
T_j	Normal force in element "i"
x_i	Coordinate of node i along X direction
θ	Relative rotation of upper and bottom polygons
d_{ij}	Distance between nodes "i" and "j"
i	First end of element
j	Second end of element
L_{ij}	Fabrication length of element "i-j" to be inserted between nodes "i" and "j"
N_c	Normal force in cable
N_s	Normal force in strut

Notations for active control

Δt	Discretisation time
$^{\tau}x$	Displacement vector

${}^{t}\left(\frac{\partial x}{\partial t}\right)$	Velocity vector
${}^{t}\left(\frac{\partial^2 x}{\partial t^2}\right)$	Acceleration vector
Δx	Displacement increment vector
M	Mass matrix
C	Damping matrix
${}^{t}K$	Tangent stiffness matrix
${}^{t}F$	Internal forces vector
${}^{t}F_c$	Actuator force
${}^{t}F_{ext}$	External loads vector
${}^{t}B$	Matrix related to the position of actuators
Q_1, Q_2, R	Ponderation matrices
a_i, i=1,...,6	Newmark's coefficients

Chapter 6

φ	Incremental parameter
[A]	Equilibrium matrix
[B]	Compatibility matrix
$[B]^-$	Generalised inverse of [B]
[D]	Displacement mode matrix
[F]	Flexibility matrix
[I]	Identity matrix
[S]	Self-stress mode matrix
[ΔB]	Increment of [B]
{d}	Displacement vector
$\{d_e\}$,	Elementary displacement vector, which generates elastic deformations in the members of the structure
$\{d_j\}$	Displacement vector of end "j"
$\{d_k\}$	Displacement vector of end "k"
$\{d_r\}$	Elementary displacement vector corresponding to rigid body displacements
{e}	Elastic deformation vector
{f}	External actions vector
{T}	Internal forces
$\{x_j\}$	Coordinate vector of end "j"
$\{x_k\}$	Coordinate vector of end "k"

$\{\alpha\},\{\beta\}$	Vectors of arbitrary constants
$\{\delta d\}$	Virtual displacement
$\{\delta e\}$	Elongations associated with virtual displacement
A	Cross section area
c_i	Cable length "i"
$C_{ij,k}$	Circle with centre k and radius ij
E	Young modulus
L	Element length
L_b	Length of the strut (geometrical model)
r_A	Rank of equilibrium matrix
R^b	Member space
r_i	Ratio strut length/ cable length
R^n	Node space
s_i	Strut length "i"
$S_{ij,k}$	Sphere with centre k and radius ij
Π	Total potential energy of the structure
$\delta^2\Pi$	Second variation of the potential energy
$\delta\Pi$	First variation of the potential energy

Main notations for contact modelling

$\{t_j^2\}$	Translation vector
$\{r_j^2\}$	Rotation vector
$\{{}^{\cdot}d_j^2\}$	Displacement vector
(Γ_1)	Final configuration
(Γ_o)	Initial configuration
$\{\delta_1\}$, $\{\delta_2\}$, $\{\delta_3\}$, $\{\delta_4\}$	Directions of translation

Preface I

Tensegrity structures are the most recent addition to the array of systems available to the designer. the concept itself is about eighty years old, and it came not from within the construction industry, but from the world of arts! Although its basic building blocks are very simple – a compression element and a tension element – the manner in which they are assembled in a complete, stable system is by no means obvious. It is also not intuitively obvious how a tensegrity system transfers loads. This stands in marked contrast to such structures as, say, suspension bridges, where the mechanism of the load transfer can be immediately grasped by even small children.

Perhaps because of these conceptual difficulties, progress in the realization of tensegrity structures has been rather slow. Apart from the tower, until very recently the one notable field of application was the tensegrity dome, a number of which are in existence.

The author of this book, René Motro, is widely recognized as one of the pre-eminent experts in the field, to which he devoted much of his career. I am convinced that this volume will go a long way toward making the concept, the theory and the practicalities of tensegrity much more accessible. As always, those willing to devote significant energies to the task will reap rich rewards. the design professionals will be able to design better structures. Interested non-professionals will experience the great pleasure of being able to say "I understand why the Hisshorn tower stands up".

I thank Professor Motro for writing the book, and I wish the readers many happy hours contemplating the secrets of tensegrity.

Stefan J. Medwadowski
Past President of the
International Association for Shell and Spatial Structures

Preface II

René Motro's "Tensegrity" is the most elaborate and most comprehensible publication on tensegrity structures I have read. Many of the books on tensegrity or tensegric structures deal with this structural system simply from the viewpoint of geometry and structural mechanics. But Motro goes deeper to understand the structural system, by investigating first the definition of tensegrity with the widest applicability. According to Motro "A tensegrity system is a system in a stable self equilibrated state comprising a discontinuous set of compressed components inside a continuum of tensioned components". Starting from the well-known description by Richard Buckminster Fuller of tensegrity as "islands of compression inside an ocean of tension", he deliberately tried to define the system in the most rational way, and reached the "extended definition " shown above. In his definition the term "inside" is a key word, and all the components on the boundary surface should be tensioned members. Thus what is called a "cable dome" as adopted by David Geiger and Matthys Levy is excluded from tensegrity according to the definition, as it has the boundary members in compression, although its structural effectiveness is recognized.

A unique idea of a balloon analogy is introduced when he tries to explain such fundamental concepts as prestress and selfstress states, formfinding, infinitesimal and finite mechanisms. Readers can understand those important concepts more easily with introduction of the balloon analogy.

Foldable tensegrities is a topic unique to this book, since it is a result of the author's study for more than ten years. the information of this chapter may be helpful in research of deployable structures. In the final chapter on Actuality of Tensegrity he confirms that tensegrity is now applicable to architecture as an established structural system, while it can be applied to other fields as well.

I am sure the reader will benefit very much from this book in terms of a profound understanding of tensegrity and of grasping the general nature of structural systems.

Mamoru Kawaguchi
Professor, Hosei University
President, International Association for Shell and Spatial Structures, 2002

Acknowledgements

This part of the book is certainly the most difficult to write, since it is difficult for personal feelings not to come into play. For me, the story, the story of the study of tensegrity systems began with David Georges Emmerich's publications that I discovered back in the September of 1968. Two months prior to that date I completed my engineering degree and embarked upon architectural studies. His so-called handbook on "*Géométrie constructive*" (Constructive Geometry), published one year previously, was for me the equivalent of a life-line after so many years of equations and rationality. Without further delay, I was soon struggling with three metallic tubes and some meters of rope, attempting to carry out the most basic of tensegrity systems. In common with many others I tried to understand this unique constructive system. Even if my relations with David Georges Emmerich were no as good as I might have wished, I recall very vividly a few hours of dialogue when we were together before an examination board. On this occasion he explained to me why he called these systems "autotendants"; it was a kind of "*risqué*" joke with his students. Ten years ago I invited him to join a structural morphology seminar, which took place in Montpellier in the September of 1992. His role was that of chairman during the workshop on tensegrity. For all these reasons, his name sprang to mind when I began to write these few words of acknowledgement.

The first time I met Stéphane Du Chateau was in 1969. He was giving a lecture on spatial structures at the School of Architecture in Montpellier. From that time until his death some thirty years later, we worked together on numerous occasions. and it was indeed an honour for me to write a few paragraphs relating to his biography for the exhibition "*L'art de l'Ingénieur, constructeur, entrepreneur, inventeur*" which was organised in Paris during the summer of 1997.

My earliest research was devoted to spatial structures, then to membranes and tensegrity systems. Throughout these years, I had the opportunity to meet, to discuss and to be enriched by many people, mainly in the Research Centre of Space Structures, in Guildford, UK. Stéphane Du Chateau introduced me to this Research Centre, and Zigmund Stanislas Makowski provided me with numerous opportunities to understand space structures. He would always do whatever he could to help me. What would be the best possible solution? It would be to work in Guildford with Hoshyar Nooshin. I have been shuttling regularly between Montpellier and

Guildford since 1973, and, unless I am mistaken, my feeling is that we have certainly become the closest of friends.

"The development of form" was the title of the first International Association for Shell and Spatial Structures (IASS) symposium where I delivered a lecture in Morgantown (1978). This was my first opportunity to meet and listen to Heinz Isler who shared his great passion for shells with all those attending. Listening to him speak has always been an inspirational experience, even if he refers to "cushions" as a source of inspiration. IASS symposia have always been most rewarding for me. It was President Steve Medwadowski who encouraged me to write this book, and provided me with the possibility of presenting tensegrity systems during IASS events. He was also very enthusiastic, when together with Jean François Gabriel, Ture Wester and Pieter Huybers, we suggested the creation of a new IASS working group called "Structural Morphology Group". It was at this moment that the so-called "gang of four" was born. It grew rapidly to become a very active working group.

I have always been impressed by Mamoru Kawaguchi's lectures: all is presented in simple terms, or, more precisely everything seems to be so simple when he explains the basic structural principles that he applies in his engineering activity.

I cannot forget the role of Sergio Pellegrino, who always has a pertinent question to ask and numerous explanations to give. It is also my pleasure to thank Ariel Hanaor: his own studies on tensegrity are very important and our scientific exchanges most fruitful. I met Maurice Lemaire during our engineering student days; it is an appropriate moment, therefore, to acknowledge our close personal friendship and to thank him for his advice and his confidence in me.

Last but not least, a special mention for Kenneth Snelson. He invited me to his studio, back in 1994, and I listened to him with much pleasure as he spoke about tensegrity.

My research has always been carried out in two places: the "Laboratoire de Mécanique et Génie Civil" (University Montpellier II), and the research group "Structures Légères pour l'Architecture" (Ecole d'Architecture Languedoc Roussillon). I have worked with numerous people and it is not possible for me here to classify each and every aspect of this co-operation. What I do know however is that everyone has been very important in helping me to understand what I have tried to explain in this book. No one objected when I asked to use our joint research and for this I am grateful. the simplest way of thanking them is to produce the impressive list below. Some were particularly involved in specific parts of this book, their names being quoted in reference lists. Others are not, since their help was of a different nature. What I can express here is the sentiment that they are very much a part of my consciousness. Of course these people could have been listed chronologically but I agree with René Daumal when he writes:

"the last step depends on the first, but the first depends on the last"

I agree with him because every single step I have taken is indeed dependent on another.

Autuori B.
Averseng J.
Belkacem S.
Bouderbala M.
Brakchi R.
Brochiero E.
Ballouard P.
Bonnet Causse R.
Carrère C.
Clary A.
Cevaer F.
Crosnier B.
Debeaud S.
De Gobbi E.
Denaeyer G
Djouadi S.
Foucher O.
Kazi N.
Kebiche K.
Grandjean A.
Hammadi H.
Joubert A.
Laporte R.
Le Saux C.
Martin V.
Marty A.
Maurin B.
Mohri F.
Moussa B.
Najari S.
Noël O.
Pauli N.
Pons J.C.
Quirant J.
Raducanu V.
Rampon A.
Sanchez R.
Smaili A.
Tuset J.
Vassart N.

René Motro

"The world is a harmony of tensions"
(Heraclites of Ephesus)

1

Introduction

"Wonder is the first step to knowledge"
(W.J. Emerson)

More often than not, "surprise" and "fascination" are the words, which convey the reactions of people who discover tensegrity systems. It is through consulting the work of D.G. Emmerich – entitled "Constructive Geometry" [Ref 1-1] – that my interest in "*systèmes autotendants*" (as Emmerich referred to them) was born. The paucity of available literature on this subject, apart from that relating to patent texts, prompted me to develop research in this area. There did indeed exist a paper by this very same author which was published in the proceedings of the first International Conference on Space Structures[1]. But it provided no clear answer to the fundamental question of stability that these systems produced. In this text David Georges Emmerich presented several illustrations, which he referred to as "Divers equilibrium et réseaux autotendants" (Figure 1.1) [Ref 1-2].

Figure 1.1 *"Equilibrium"*

[1] This conference was organised in 1966 by Z.S. Makowski, at the Space Structures Research Centre (Guildford, UK). Three others have followed it in 1975, 1984 and 1993. These conferences had a major impact mainly among architects and engineers.

How can these assemblies be stable, since their heaviest elements seem to float in space and their lightest elements appear to escape the eye? Is there some secret involved here? The reply to this last question, needless to say, is negative. Simple explanations can be and indeed are provided for the stability of tensegrity systems. But David Georges Emmerich's text did not provide such mechanical clarification. In 1978, some ten years after its publication, my interest in the subject took a new dimension during a stay in Washington, when, having admired the challenge to equilibrium of Calder mobiles (Figure 1.2) in the Museum of Modern Art renovated by I.M. Pei, I walked into the Hisshorn Museum.

Figure 1.2 *Mobile by Calder*

Within the walls of its garden, I discovered first with surprise and then fascination K. Snelson's works. And, in particular, a sculpture some thirty meters high, the aesthetics, purity and rhythm of which could not fail to surprise the visitor: it was a mast named "Needle Tower" (Figure 1.3).

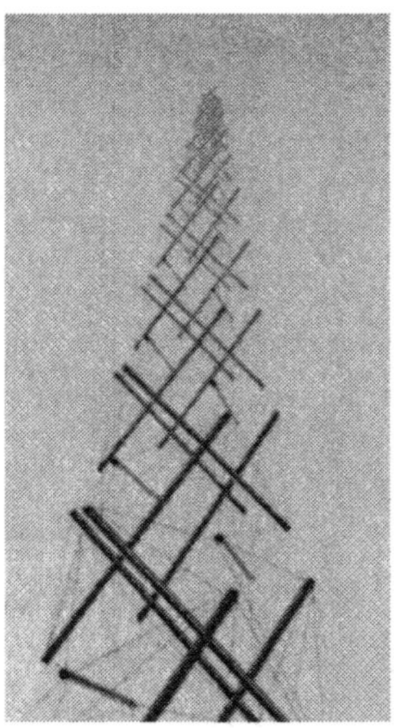

Figure 1.3 *Needle Tower*

How did it work? From a purely geometrical point of view, the multiplication of cells with identical composition (but of different size), organised the system around a vertical axis. The geometrical composition became obvious; when placing myself opportunely at the mast's basis my eye discovered ordaining symmetries (Figure 1.4): it was at that moment that I came to understand and have since gone on to explain its structural composition [Ref 1-3].

Figure 1.4 *Needle Tower organising geometry*

But many other questions remained unanswered. What was the static principle of this sculpture? How was it completed? What was the principle of its structural composition? Equally important – what sequence of events had led Snelson to achieve this sculpture? Did this tower, whose top one could not perceive, have any particular symbolism? The meaning of this work, or at least one of the approaches I could make, related with this duality traction–compression and to this verticality that came back to me several years later, when I read the "Pendulum of Foucault" written by Umberto Eco [Ref 1-4]. Elsewhere on Washington Mall there was a science museum where a pendulum was exhibited. The question was and remains the search for a handling point in the sky for the pendulum. But for all its obviousness there was no such point... and the question of stability would thus remain unresolved.

All people who have been surprised and fascinated when seeing tensegrity systems have not escaped these very same questions. Some of these have a reply, mainly mechanical interrogations. The object of the "Fundamental concepts" chapter is to give a view about fundamental mechanical concepts which these systems obey; consequently it offers the possibility for the reader to create models for himself that will help him to confront his thoughts, through direct and physical contact with tensegrity systems.

One of the best introductions to tensegrity systems and therefore to this book is certainly to build a small model. Effective realisation always appears useful; it is possible, for example, to simply construct "elementary equilibrium". It will be composed of three compressed elements of equal length and nine tensioned elements of equal length (Figure 1.5).

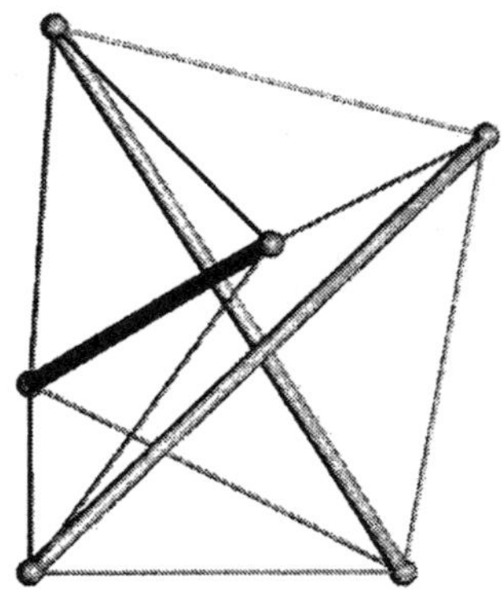

Figure 1.5 *Elementary equilibrium*

This equality of lengths for each of the two classes of elements justifies the qualification "regular" that we will adopt for this elementary system. It is necessary in this specific case that the ratio "r" between the length of compressed elements "s" and that of tensioned elements "c" be equal to 1.468[2]. A different choice of these values creates either a system without rigidity, a system to which one cannot give a shape ("r" is too small), or a system which will be very difficult and perhaps even impossible to assemble ("r" too great).

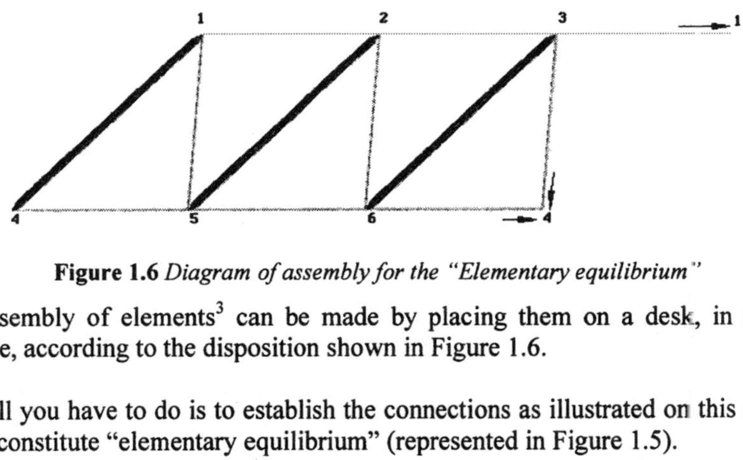

Figure 1.6 *Diagram of assembly for the "Elementary equilibrium"*

The assembly of elements[3] can be made by placing them on a desk, in the first instance, according to the disposition shown in Figure 1.6.

Then all you have to do is to establish the connections as illustrated on this diagram and to constitute "elementary equilibrium" (represented in Figure 1.5).

---

[2] This value results from a "form-finding" process, which will be explained in subsequent chapters.

[3] Wooden rods with small nails at their ends and fishing line are well adapted for this purpose.

This is the first step on the road to knowledge concerning tensegrity systems. Keep hold of your first own tensegrity system; it will open the keys in respect of the following chapters!

There are six chapters. The first *"History and Definitions"* is the opportunity to refer to the pioneers of tensegrity and to discuss possible definitions. The concept of tensegrity can be applied outside the simple scope of architecture and engineering, and we submit a definition that can be applied for material as well as for non-material systems (like sociological systems).

When tensegrity is applied to material systems, it is useful to describe the "*Fundamental Concepts*": relational structure, forms and forces are discussed in this chapter. Coupling between forms and forces is omnipresent in these systems with initial stresses. Associate self-stress states and mechanisms are illustrated with simple examples.

Even if tensegrity systems are simultaneously characterised by mechanical and geometrical features, "*Typologies*" have first been established with regard to their shape: cells and assemblies of cells are described in this chapter that has to be considered as a historical classification of typologies.

How can tensegrity systems be studied? Which design process do we need? The reply is, of course, that there is no single solution. Several "*Models*" are available: depending on the problem to solve, one or the other might be more appropriate, sometimes several have to be used simultaneously. Even if this book is not a theoretical book, the reader will find in this chapter some thoughts relating to these models, and also references to scientific literature. We have tried to give a comprehensive approach of form-finding – which is certainly the main problem to solve: experimental and theoretical design processes are studied. Self-stress and related questions are also raised in this chapter, the last paragraph of which is devoted to active control of tensegrity leading to smart structures.

Mechanisms and self-stress states are an asset of tensegrity systems; they allow us to design "*Foldable Tensegrities*", using alternatively activation of finite mechanisms in order to fold the system and the stiffening effect of self-stress to deploy them. Some design examples are described in this chapter, the last part dealing with theoretical problems which occur during the modification of shapes.

Finally, aware that we are only at the beginning of the use of tensegrity systems, we refer under the heading "*Tensegrity: latest and future developments*" to some current projects: that is to say new tensegrity grids and other projects under completion. The future of tensegrity now has to be written; considering tensegrity systems as a structural principle will certainly be fruitful, avoiding the belief that everything can be reduced to architecture and engineering.

References

Ref 1-1. Emmerich D.G., Cours de Géométrie Constructive – Morphologie, École Nationale Supérieure des Beaux Arts, Paris, Centre de diffusion de La Grande Masse, Paris, 1967.

Ref 1-2. Emmerich D.G., "Réseaux", in Space Structures: A study of methods and developments in three-dimensional construction, proceedings of the International Conference on Space Structures, Guildford, 1966, edited by R.M. Davies, Blackwell Scientific Publications, 1967, pp. 1059–1072.

Ref 1-3. Contemporary Developments in Design Sciences, exhibition, Cathedral of St John the Divine, New York, USA, November, 1995.

Ref 1-4. Eco U., Le Pendule de Foucault, © Gruppo Editoriale Fabbri, Bompiani, Songzono, Etas S.pA. Milan, 1988, © Editions Grasset et Fasquelle, 1990.

2

History and Definitions

2-1. History

2-1.1. Ioganson and constructivism

D.G. Emmerich has reported what appeared to him to have been the first structure that can be placed in the proto-tensegrity system category [Ref 2-1]. He refers to the research carried out by the Russian constructivists, which is described in a book by Laszlo Moholy Nagy, "Von Materiel zu Architektur", first published in 1929 and republished in 1968. Laszlo Moholy Nagy included two photographs of an exhibition held in Moscow in 1921 showing an equilibrium structure (Gleichgewichtkonstruktion) by a certain Ioganson.

A recent exhibition held at the Guggenheim Museum, "The Russian and Soviet Avant-Garde, 1915–1932", gives more details on the work of constructivists who organised their first exhibition manifestation, the Obmokhu (the Society of Young Artists) in May 1921. Rodchenko, one of the constructivists, claimed in January 1921 [Ref 2-2]:
"All new approaches to art arise from technology and engineering and move towards organisation and construction".

One year after Karl Ioganson wrote:
"From painting to sculpture, from sculpture to construction, from construction to technology and invention – this is my chosen path, and will surely be the ultimate goal of every revolutionary artist....".

In the exhibition mentioned above, Ioganson displayed a "sculpture-structure" completed during 1920 (Figure 2.1).

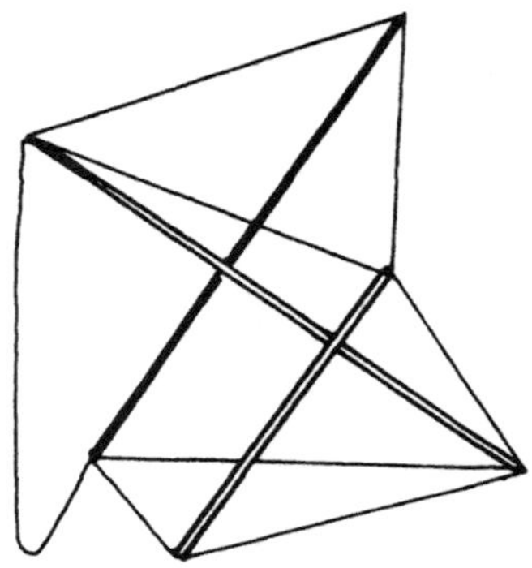

Figure 2.1 *Ioganson sculpture*

As Emmerich put it:
"This curious structure consists of three bars and seven cables and is handled by means of an eighth unstressed cable, the whole being deformable".

Moholy Nagy illustrated this structure as a "Study in Balance", explaining "...that if the string was pulled, the composition would change to another position and configuration, while maintaining its equilibrium".

"The similarity between the manner of jointing in Study in Balance and that of other constructions by Ioganson suggests that all the works could be adjusted. He was exploring the movement of skeletal geometric structures in a more pragmatically experimental and explicitly technical manner than was Rodchenko in his hanging constructions. Ioganson's works do not evoke any specific structure, yet the use of standardised elements and the emphasis on the transformation of form might appear to have more direct application to utilitarian structures such as portable, fold-up kiosks or collapsible items of furniture."

This transformable structure was the result of a very clever structural design, even if it doesn't relate to any kind of pre-stress or self-stress, which characterise tensegrity systems. In fact, recently, a similar prototype has been completed for research purposes, to investigate the foldability of tensegrity systems [Ref 2-3]. The various states of static equilibrium can be understood as funicular states and Ioganson's construction is very well adapted to explain mechanisms. According to structural morphology it illustrates the fact that several shapes can be linked to a single structure (this word being understood in its relational meaning).

2-1.2. Concept, words and design

Many works have been devoted to the history of tensegrity systems. The two main references are contained in special issues of the International Journal of Space Structures, published in 1992 [Ref 2-4], and in 1996 respectively [Ref 2-5]. One will bear in mind that, as always, a controversy exists between three people, namely

David Georges Emmerich, Richard Buckminster Fuller and Kenneth Snelson. It might not have escaped your attention that, as a precaution, I have named them in alphabetical order! All three have applied for patents that give ample testimony to faith in the subject, at least in legal terms. It is necessary to note, and this is important, that all three protaganists described identical structures, deriving from a module comprising three struts and nine cables.

Generally, so far as a concept is concerned, one can not define it completely. Nevertheless we can illustrate it – and my memory retains mainly the idea of Richard Buckminster Fuller describing tensegrity systems as *"islands of compression in an ocean of tension"*. as is quoted in the International Journal of Space Structures (1996 special issue):

Fuller wrote in "Designing a New Industry", a booklet published by the Fuller Research Foundation, Wichita, Kansas, 1945–46:

"We find in the mechanical structuring of the universe, that compressive organisation is limited to the dimensional confines of heavenly spheres themselves, and that vaster structural integrity of the universe is maintained within the infinite limits of tensile stress principles only, which we identify as gravitational attraction. This is truth, I am going to pursue this truth into demonstrated technical advantage by man. These are principles I must employ in a big way in putting environment under man's direct control."

In a talk given by Fuller at the University of Michigan Mid Century Conference on Housing, in April, 1949 and published in 1963 in Ideas and Integrities, he is a little more specific:

"...Tension is comprehensive. Universe tensionally coheres non-simultaneous events.
...Universe is tensional integrity".[1]

Indisputably, Fuller was the promoter of the concept of "tensional integrity", even if his writings are difficult to understand line by line. When this concept is applied to structural systems, many of them fit, and one can for example claim that a balloon and more generally an inflated membrane conforms satisfactorily to Fuller's idea, since no precision of matter or shape is given in his approach. On the other hand, indications are provided on isolation of a matter in a state of compression immersed in matter in a state of tension: air is compressed inside a tensioned envelope.

[1] It would be wrong to read Fuller's texts as scientific, since some of his propositions are not sufficiently explicit from a scientific point of view. But his texts can be considered as visionary since currently some people such as Donald Ingber use the tensegrity principle in biology.

If we are indebted to Fuller for the concept, in the author's opinion the birth of its application to space structures comprising struts and cables seems to be the result of Snelson's work.

The word "Tensegrity" itself results from the natural contraction of "tensional" and "integrity". Fuller made this contraction as he did for the three words "Dynamic", "Ion" and "Maximum" to create the famous term "Dymaxion", which qualified many of his projects.

2-1.3. Patents

2-13.1. Chronology

Several documents show that R. Buckminster Fuller applied for patents in the USA almost simultaneously with D.G. Emmerich in France, at the very beginning of the 1960s. R. Maculet [Ref 2-6] gives the list of patents taken out by D.G. Emmerich and his comments on the reception of a patent concerning "frame assembly elements, in particular for the building industry..." (June 1959). The first patent referring to self-stressing[2] systems is dated on 1963.

R. Maculet found four inventions published by Buckminster Fuller concerning tensegrity systems, the oldest (1962), in a work dated 1985 with no author's name [Ref 2-7]. The same date is mentioned in the journal Synergetica [Ref 2-8] concerning a patent application by Gwilliam et al. The numerous patents that he has mentioned on this occasion showed the increase in patent applications in the preceding years.

Emmerich and Fuller patents were applied for between 1959 and 1964. Snelson's is dated 1965.

If we look to chronology, no precise conclusion can be reached. David Georges Emmerich patented a first system called "Pearl Frameworks " at the INPI (Institut National de la Propriété Industrielle) [Ref 2-9], but it was not correctly registered. His second patent is dated 1963 and granted in 1964 and was entitled "*Construction de réseaux autotendants*" (Figure 2.2). Fuller's main patent was registered in 1959 and granted in 1962 (Figure 2.3). Fuller chose the name "Tensile Integrity" [Ref 2-7]. Snelson's patent "Continuous tension, discontinuous compression structures" was registered in 1960 and granted in 1965 (Figure 2.4).

---

[2] The word "self-stressing" is the translation of the word "autotendants" in French.

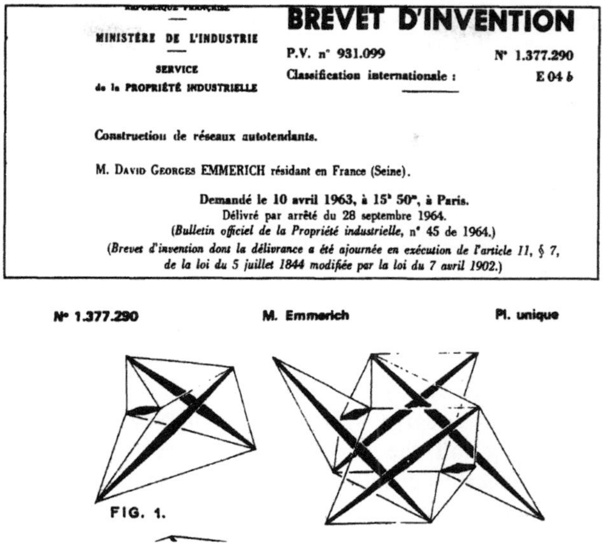

Figure 2.2 *Emmerich's patent*

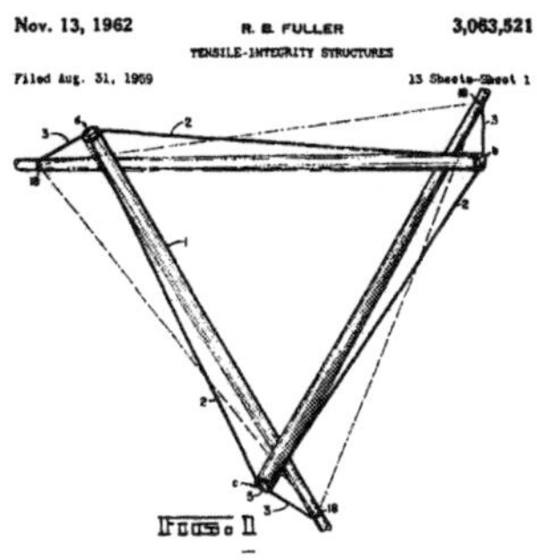

Figure 2.3 *Fuller's patent*

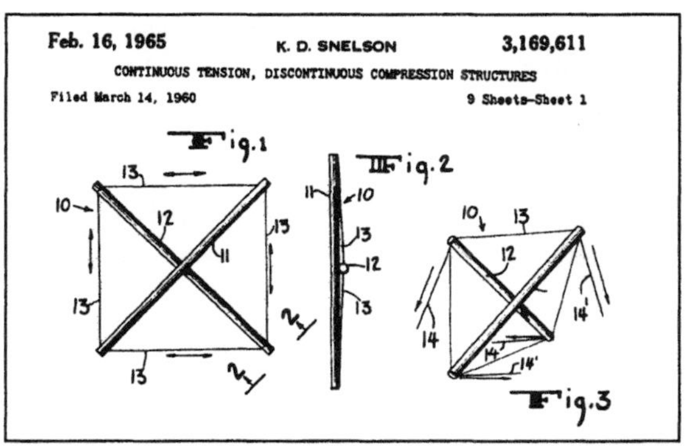

Figure 2.4 *Snelson's patent*

What happened thereafter is a source of controversy between Snelson and Fuller. Furthermore, David Georges Emmerich in France, claimed that he was the inventor of this new structural system, which he referred to as "autotendants".

Readers interested in a more precise description of these controversies may refer to papers published in 1996 in the *International Journal of Space Structures*. Two points may be underlined. Firstly, the structural system described by the three men is the same. Secondly, it was Fuller who created the word 'Tensegrity'.

Patents are the administrative proof for intellectual property. Searching for the earliest patents is the domain of specialist organisations; it will be more fruitful for us to look at the way which led to what is called tensegrity systems, and to understand the work carried out by several people in relation to this specific kind of structure. If a preview of tensegrity systems was achieved by Ioganson, the birth of the elementary tensegrity unit, the so-called "elementary equilibrium" (name given by D.G. Emmerich), clearly appears in Snelson's patent as a sub product of an assembly of planar units. Patents from Fuller and Emmerich are not so explicit. A precise comparative analysis of the three patents could be useful but this work has yet to be carried out, since the controversies described in the special issue of the International Journal of Space Structures in 1996 does not give a clear explanation on the initial design process. An attempt has been made last year (2001) in our laboratory and it will be detailed in Chapter 5 ("Models"). Moreover a remaining question is the link with Ioganson's sculpture: we know that Emmerich saw this sculpture. Did Ioganson also inspire Snelson? The answer is not clear. It is perhaps better not to attribute too much importance to finding out who was first but rather to examine the future of these systems.

2-13.2. "Continuous tension, discontinuous compression structures"

"*Continuous tension, discontinuous compression structures*" is the title of the patent awarded to K. Snelson (Feb. 16, 1965, Patent No 3,169,611) [Ref 2-10]. We must pay particular attention to the birth of the concept included simultaneously in this patent and in the description of the earliest works carried out by Snelson, which will be described further in this section and also in Chapter 5. It is not my aim to nominate Snelson as the father of Tensegrity Systems. I am not qualified to do such a thing, but information subsequently acquired, in my opinion, is certainly of great interest for people who wish to know more about the birth of Tensegrity systems.

Correspondence between K. Snelson and myself sheds interesting light on the respective roles of Fuller and Snelson[3]. And this was confirmed when I met K. Snelson in 1984.

Fuller defined the emergence of tensegrity as follows:
"The word tensegrity is an invention: it is a contraction of tensional integrity ... tension is omnidirectionally coherent. Tensegrity is an inherently non-redundant

[3] K. Snelson, private correspondence with the author, 1990. See Appendix C.

confluence of optimum structural-effort effectiveness factors. Tensegrity structures are pure pneumatic structures and can accomplish visibly differentiated tension–compression interfunctioning in the same manner that is accomplished by pneumatic structures, at the subvisible level of energy events... " [Ref 2-10].

Edmonson also reports that: [Ref 2-12]:
"In the summers of 1947 and 48, Fuller taught at Black mountain College and spoke constantly of tensional integrity. Nature relies on continuous tension to embrace islanded compression elements, he mused; we must create a model of this structural principle ... Much to his delight, a student and now well-known sculptor, Kenneth Snelson, provided the answer".

Let us allow Snelson to explain for himself the work he showed to Fuller:
"The three small works which are of interest here were concerned both with balance of successive modular elements hinged one-to-another and stacked vertically as seen in Figure 2.5; and later, suspended one-to-the-next by means of thread-slings as shown in Figure 2.6. One can see module-to-module sling tension members replacing the wire hinges connecting the modules shown in Figure 2.5. One step leading to next, I saw that I could make the structure even more mysterious by tying off the movement altogether, replacing the clay weights with additional tension lines to stabilise the modules one to another, which I did, making "X", kite-like modules out of plywood. Thus with forfeiting mobility, I managed to gain something even more exotic, solid elements fixed in space, one-to-another, held together only by tension members (Figure 2.7)".

Figure 2.5 *Kenneth Snelson, "One to another", 1948*

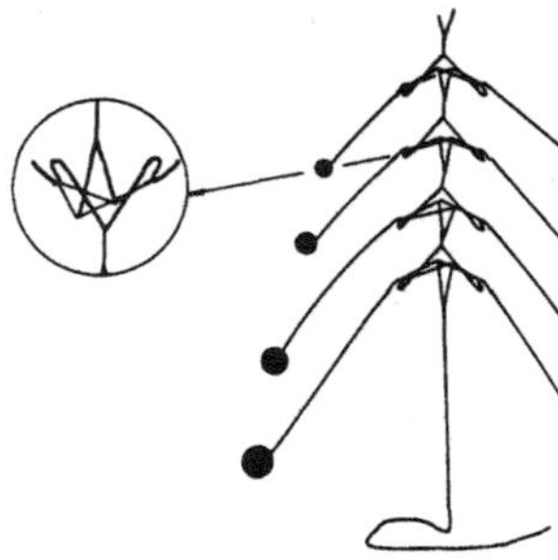

Figure 2.6 *Kenneth Snelson, "One to the next", 1948*

Figure 2.7 *Kenneth Snelson, "X-shape", 1948*

A first analysis of this explanation could be the following. The first sculpture looks like Calder's work. The key to the process lies in the evolution of junction mode between elements: the assembly mode, which is made with rigid wires in this sculpture, is replaced in the second one by four cables assembled in a spatial rhombus shape giving more degrees of freedom to the whole. An X-shape appears between these cables and emphasis is put on this part in the third sculpture (Figure 2.7), where stabilising cables are added. These cables replace the actions of the clay balls.

The last model needs to be carefully studied in relation to the explanations given by Snelson in his patent. The "X" module is the basic idea of the patent awarded to Snelson for *"improved structure of elongate members which are separately placed either in tension or in compression to form a lattice, the compression members being separated from each other and the tension members being interconnected to form a continuous tension network".*

In the same patent Snelson writes:

" the basic module disclosed by the invention utilises only two elongate compression members and an associated tension network, to form a self-supporting structure. The compression members cross each other at some intermediate point between their ends in an X-shape or a modified form thereof. The outer ends of the compression members are pulled toward adjacent ends by tension members comprising a continuous tension network. Means are provided, either in the construction or shape of the compression members themselves, or in the use of additional tension members, for separating the compression members at the points where they cross each other. "

The conceptual work described shows a morphological evolution from plane structures to spatial ones by this introduction of additional tension members (Figure 2.8). The morphological units are combined and Snelson provides detailed explanations for designing complex structures. Snelson plays with X-shape along unidirectional or multi-directional axis. One of his projects is described in the next section: it is the "key" unit that has been much used in recent years.

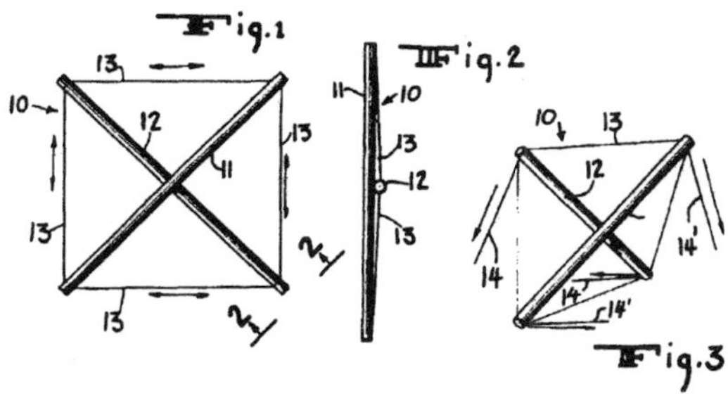

Figure 2.8 *Kenneth Snelson: Continuous Tension, Discontinuous Compression. Basic concept*

2-13.3. Basic spatial tensegrity systems

Attention must be paid to one specific tensegrity system described in Snelson's patent. Assembling three X-shape moduli, and adding necessary cables to disconnect compressed members, leads to an assembly of nine cables (Figures 2.9 and 2.10). The resulting set of cables is composed of two triangles and three bracing cables. When three compressed members are included in this set a "prismatic" unit is created. According to the pre-stress principle, this is the simplest way from plane pressurised basic units to spatial self-stressed structures (Figure 2.11). This basic unit has been called "simplex", and many authors such as D.G. Emmerich have comprehensively presented its description, but in the author's opinion its structural generation is only clearly displayed in Snelson's patent.

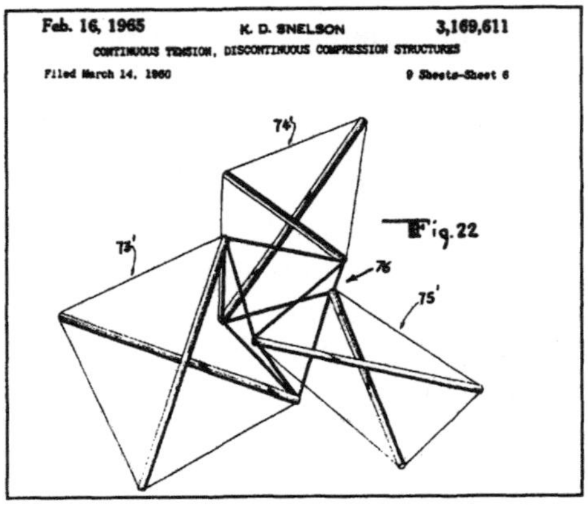

Figure 2.9 *Kenneth Snelson, Assembly of 3 X-shapes*

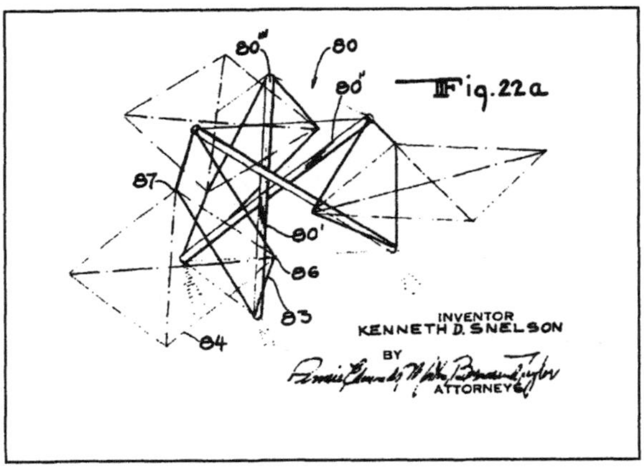

Figure 2.10 *Kenneth Snelson, Elementary spatial tensegrity system*

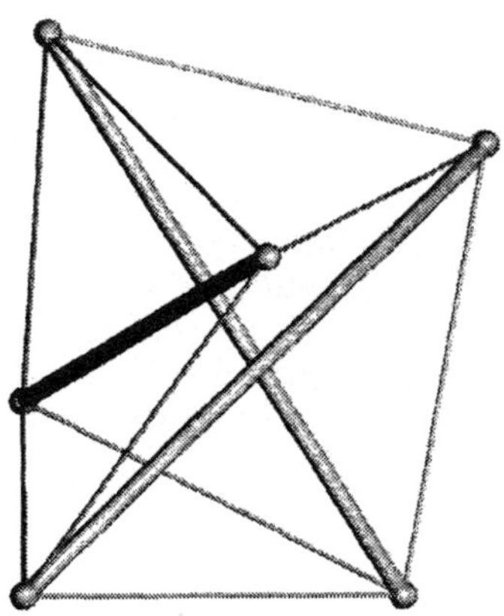

Figure 2.11 *Self-stressed space structure. Elementary equilibrium*

2-2. Definitions

2-2.1. Concept and definition(s)

If the origins of tensegrity systems remain a matter of controversy, it is also very difficult to give an unambiguous definition as previous attempts have proved[4]. Some elements concerning these proposals can be found in the literature ([Ref 2-10] [Ref 2-13], [Ref 2-14], [Ref 2-15]).

As far as the concept itself is concerned, it is usually very difficult to define it completely. Nevertheless a concept can be illustrated, and the best way is to quote Richard Buckminster Fuller describing the tensegrity principle as *"islands of compression inside an ocean of tension"*. On this simple basis many objects could be related to the tensegrity principle: a balloon and more generally any inflated envelope fits with it, since no precision on matter, or form is included in the concept. But indications are given on the isolate character of compressed matter immersed in a tensioned matter. In a balloon, the compressed air is included inside the pre-stressed envelope. In the following paragraphs we give a first definition based on patents, and discuss it. It is then possible to extend this definition while at the same keeping a close relationship with the concept itself. This extended definition comprises the first one, but is also applicable to many other systems such as biological cells.

2-2.2. First definition based on patents

The difficulties in establishing a clear definition of tensegrity systems are well-known. It seemed some years ago to be useful to give a "patent based" definition that could serve as a reference for comparison with the other known definitions. On the other hand, it could constitute a kind of reference to judge to what extent such or such a constructive system can be related to the class of tensegrity systems. To constitute this kind of reference, the following definition is established on the basis of patents, which have been registered by Fuller, Snelson and Emmerich. All three describe the same structure and it is in this meaning that the corresponding definition can be qualified as a "patent based" definition (or in abbreviated form "patent" definition) bearing in mind its relativity.

---

[4] It is not the "privilege" of tensegrity systems: I attended, in 1998, a colloquium in Cambridge on "deployable systems", and the President of the International Association of Shell and Spatial Structures, Steve Medwadowski, talked about the definition of "deployable systems". It appeared that it was very difficult to give a comprehensive definition of "deployable systems". My problem at that time was of course rather significant, since my own talk was on "deployable tensegrity systems"!

> *Tensegrity Systems are spatial reticulate systems in a state of self-stress. All their elements have a straight middle fibre and are of equivalent size. Tensioned elements have no rigidity in compression and constitute a continuous set. Compressed elements constitute a discontinuous set. Each node receives one and only one compressed element.*

Constitutive proposals of this definition call for the following comments:

1 *Tensegrity Systems are spatial reticulate systems*: this is an affirmation of the spatiality, and of a structural layout that causes in elements pure compression or tension states of stress. "Tension Systems " are a subclass of spatial reticulate systems: they comprise elements that have no rigidity in compression. These elements, and only these elements, in all circumstances, are tensioned: tensegrity systems have this characteristic and pertain to this subclass.

2 *They are in a state of self-stress*: stiffness is produced by the self-stress, independently of all external actions, connecting actions included. The self-weight is not taken in account at the design step, and it does not contribute to their initial equilibrium.

3 *All their elements have a straight middle fibre and are of equivalent size*: this third point is implicitly present in the first known patents. It is certainly one of these that caused many of the controversies, especially the mention concerning the equivalence of the elements' size.

4 *Tensioned elements have no rigidity in compression and constitute a continuous set*: these elements are generally cables. The continuity of tension set contributes to the aesthetics of tensegrity systems. On the mechanical level, their presence is often the source of misunderstanding, since designers and builders generally work on reticulate systems whose elements have simultaneously a rigidity in compression and in tension. Known results on stability, in the general meaning of this word, linked to classical reticulate systems, have to be reconsidered by taking into account this essential – and perhaps most important – characteristic.

5 *Compressed elements constitute a discontinuous set*: it could have been claimed that compressed elements do not need rigidity in tension, since all these elements are always in the same qualitative state of stress: compression. The technological execution of this condition is possible, but one rigid element in compression and in tension is generally used, even if they are never submitted to this last load effect. None of the three patents gives an indication on this point. The discontinuity of compressed elements generally stimulates a number of questions, since it constitutes, structurally speaking, a discontinuity of thought, when compared to the totality of usual constructive systems: in our unconscious, which is fed with experiences drawn in constructive archetypes, compression necessitates the continuity of transmission. Tensegrity systems throw back this way of thinking and this is mainly why they create surprise.

6 *Each node receives one and only one compressed element*: Historically speaking, systems which are described in the patents satisfy this condition. But it is also a precision, which provoked much controversy sometimes related to point 5. This precision is necessary since there can exist systems with more than one compressed element, and which satisfy the extended definition (Section 2-2.3), but some systems also exist without any compressed element for some nodes, which receive only tensioned elements. In many cases, compressed elements are struts and tensioned elements are cables. This is why in the subsequent chapter we will use "struts" in place of "compressed elements" and "cables" for "tensioned elements", when there will be no possible confusion.

2-2.3. Extended definition: tensegrity system or not?

The basic ideas are included in the concept described by the expression *"islands of compression in an ocean of tension"*. It is obvious that there are two kinds of components according to their state of load effect: compression or tension. A second character is included in the concept: *compression is inside tension.* A third idea is that compressed entities are islands: they constitute a discontinuous set. Tensioned entities are gathered in a continuous set. A last point concerns the necessary equilibrium of the whole system.

2-23.1. Definition

When compared with these characteristics, the definition given by A. Pugh [Ref 2-13] seems to be very well adapted to constitute a valuable basis for an extended definition:

"A tensegrity system is established when a set of discontinuous compression components interacts with a set of continuous tensile components to define a stable volume in space."

But it needs to be slightly modified to take into account the following factors:

- Components in compression are included **inside** the set of components in tension.
- Stability of the system is **self-equilibrium** stability.

Furthermore, some expressions like "compression components", "tensile components" and "volume" are not very well adapted.

So, I would venture to suggest the following definition:

"A tensegrity system is a system in a stable self-equilibrated state comprising a discontinuous set of compressed components inside a continuum of tensioned components."

2-23.2. Discussion

2-232.1 System

We use the word “system” in relation to the theory of systems, which has been developed to describe ordered entities. It is useful since it allows us to distinguish between:

- Components (with qualitative characteristics, and sometimes quantitative characteristics); two classes are identified according to the nature of stresses.
- Relational structure, which gives a clear description of relationships between components. It can be described using graph theory. For tensegrity systems that are homeomorphic to a sphere, the graph of tensioned components is planar (see Ref 2-16).
- Total structure that associates relational structure with qualitative and quantitative characteristics.
- Form considered as projection of the system on to a three dimensioned reference system.

2-232.2 Self equilibrium and stability

This expression expresses the initial mechanical state of the system, before any loading, even gravitational. The system has to be in a self-equilibrium, which could be equivalent to a self-stress state, with any self-stress level. It could also be said that the system has no finite mechanism. It is possible to find infinitesimal mechanisms in tensegrity systems [Ref 2-17], but these are generally stabilised by the self-stress states. Stability is defined according to the ability of the system to re-establish its equilibrium position after a perturbation. These concepts will be further developed in Chapter 3.

2-232.3 Components

Pugh used the word “component”, and it is necessary to keep it, and to avoid some terms such as “elements”, which are ambiguous. The shape of the component is not prescribed to be a line, a surface or a volume: it can be a strut, a cable, a piece of membrane, or an air volume (Figure 2.12). It can be a combination of one or several elementary components that are assembled in a higher order component. The matter of the component is not prescribed: air, steel, and composite, etc.

Figure 2.12 *Tensegrity system by Fuller*

2-232.4 Compression and tension

"Continuous Tension, Discontinuous Compression" is an expression used by Kenneth Snelson in the title of his patent. Compression and tension are mechanically speaking a load effect, which implies that the matter of one component is subjected either to a compressive or a tensile effect. Consequently a component, which is compressed, requires rigidity in compression; a component, which is in tension, requires rigidity in tension. It is only sufficient to have unilateral rigidity, compression or tension rigidity. It is very well know that cables and membranes have no rigidity in compression. In building systems, tensioned components are generally thinner than compressed components, which can be subjected to buckling phenomena.

Technologically, it is possible to find components (in association with an appropriate material) with compression rigidity and no tension rigidity (two associated tubes for instance, one inside the other with the possibility to move relatively according to a single way).

It is why we prefer to use the phrases "compressed components" and "tensioned components" in place of "compression components" and "tensile components". Even if a component is very complex in terms of shape, or if it is the result of an assembly of identifiable elementary components, the condition is that all its matter has to be compressed or tensioned according to its class.

2-232.5 Discontinuous set, continuum

These words are closely related to words "*islands*" and "*ocean*" used by Buckminster Fuller. Each compressed component constitutes an "*island*"; when a structural system is defined as a tensegrity system, it is necessary to identify each compressed component. If there is more than one, these components have to be disconnected. If we used graph theory, their associated graph would be disconnected. Systems with only one compressed component constitute a specific case and can be also considered in the scope of tensegrity systems.

A discussion could be embarked upon tensioned components since their corresponding set has to be continuous and consequently the whole tensioned components could be defined as a higher order component. There are several "oceans" in our earth, but it seems that using the expression "continuum of tensioned components" is adapted to our objective and does not cause any controversy. It is always possible to identify lower order components inside the continuum.

2-232.6 "Inside"

"*Inside*" is a key word in the definition since it will allow us to separate two kinds of structural design: one which is a part of our constructive culture and based on compression as in the sustaining load effect, and an opposite one based on tension as fundamental "support". In order to know if "*islands*" of compression are, or are not, inside an "*ocean of tension*" it is necessary to establish a clear definition of the limit, of the frontier between the inside and the outside of the system.

Problems related to topology may arise with some systems such as toruses, but it is always possible to define the inside and the outside of a closed envelope. Every system can be described by a set of nodes and modelled by points, as is done in numerical methods such as the finite element method. It is more sophisticated for continuous components, but certainly can be done. Mathematically speaking, any set of nodes admits a frontier which is generally a polyhedral convex surface, comprising triangles built with some of the nodes of the system. But this does not exclude other surfaces such as the torus, since it also possible in this case to know if a point (and the associated node) is inside the envelope or not.

Consequently, a first proposal could be that a component is considered to be inside this envelope, if all its describing points are not on this frontier envelope, and if none of these describing points are outside this envelope. This first proposal is generally sufficient, but for some cases it might not be. If we want to have a more precise definition, we need to consider the compression lines: the ends of compressed components belong to the continuum of tension, whether one of them is, or is not, on the boundary. If all the points lying on the segment defined by these two points are inside the envelope, the compressed component can be considered to be inside the envelope.

The problem is not to define only the envelope, but to look for an envelope satisfying the criteria as previously mentioned: no compression line on this envelope, only tension lines. Therefore this envelope encloses a tensegrity system and is a part of this system.

A direct consequence of this definition is that all the points on this envelope belong to the continuum of tensioned components. All the action lines lying on the boundary surface between the outside and the inside of the system are tension lines. Of course, this is the crucial point of the definition of tensegrity systems, since the self-equilibrium is based on tension and this is not usual when it comes to the history of constructions. This fact is certainly at the basis of that referred to as "*surprise*" and "*fascination*". Moreover duality between traction and compression is not complete, since the stability of self-equilibrium is not ensured when compression is replaced by traction, and vice versa, as Vassart has demonstrated so effectively. [Ref 2-17].

2-23.3. Examples

Related to our proposal it is interesting to examine some cases and to analyse them through the filter of our definition.

2-233.1 Elementary cells

In the case of elementary cells such as those that have been largely examined previously (simplex with three-strut, or six-strut system, known also as "expanded octahedron"), the two definitions, "patent" one and, "extended" one are working. The components are the struts; these systems are self-equilibrated (their geometry is a self-stress geometry). The set of compressed components is characterised by a disconnected graph (see [Ref 2-16]), and it is inside the continuum of tensioned components comprising, in this case, cables. We illustrate with Figure 2.14 the polyhedral convex envelope for the two examples mentioned (Figure 2.13).

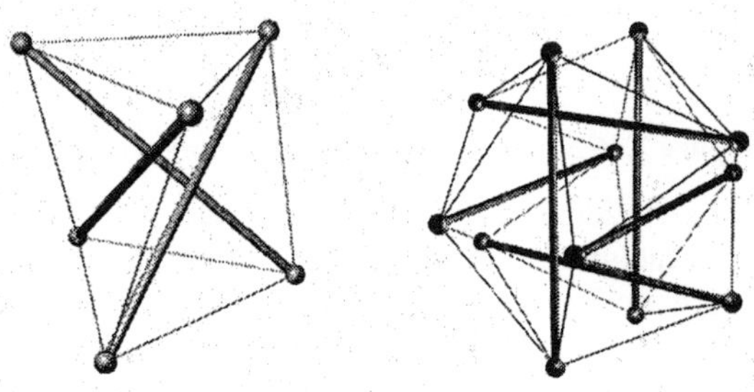

Figure 2.13 *Three-strut and six-strut tensegrity cells*

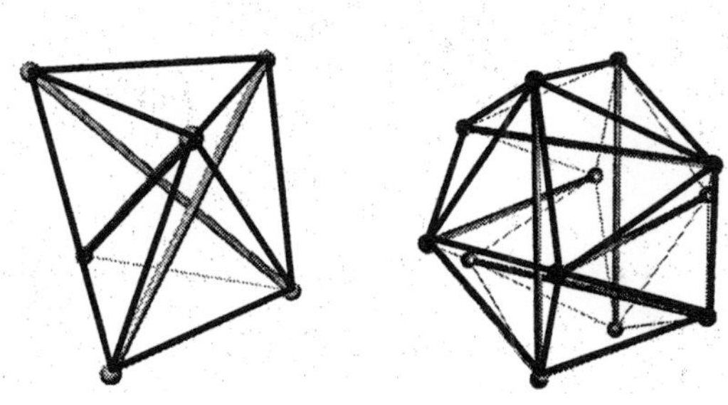

Figure 2.14 *Three-strut and six-strut tensegrity polyhedral envelopes*

The cells are such that the external set of tensile components is homeomorphic to a sphere. Consequently we can call them single tensile layer tensegrity systems. They follow a scheme which looks like a balloon (Figure 2.15): but in this case the external envelope is made up of straight tensile members which constitute a discrete net. External thin lines are under tension, internal bold lines with arrows model compression forces. The diagram is plane for the purposes of simplicity, but as in Figure 2.16, it is in fact spatial.

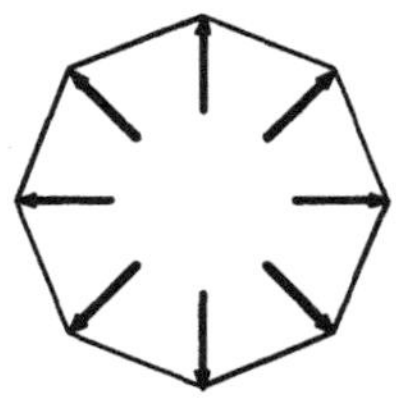

Figure 2.15 *Structural principle for a single layer tensegrity system*

This structural principle is valid for many tensegrity cells, which have been described previously. But in general cases the structural principle can be illustrated by the diagram given in Figure 2.16, where some tension lines are also included inside the external envelope.

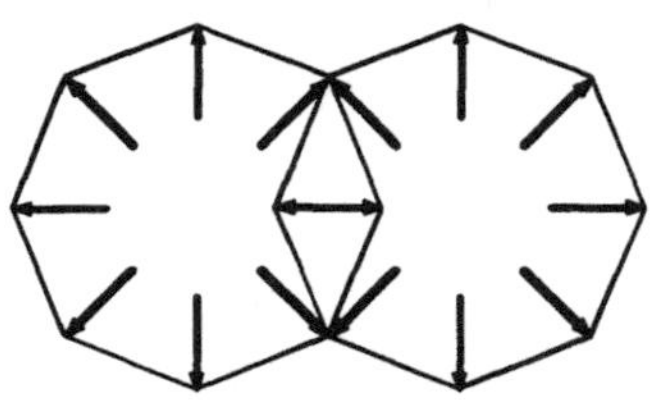

Figure 2.16 *Structural principle for tensegrity system*

It appears on this basis that tensegrity systems can be understood as discrete "pneumatic" tensile structures[5], opening thus a wide range of possibilities also allowing zero curvature, positive curvature and mixed curvature in the same system. The double layer grid of Figure 2.19 is based on flat tensegrity principle as shown in Figure 2.17.

---

[5] R.B. Fuller was aware of the pneumatic character of tensegrity system.

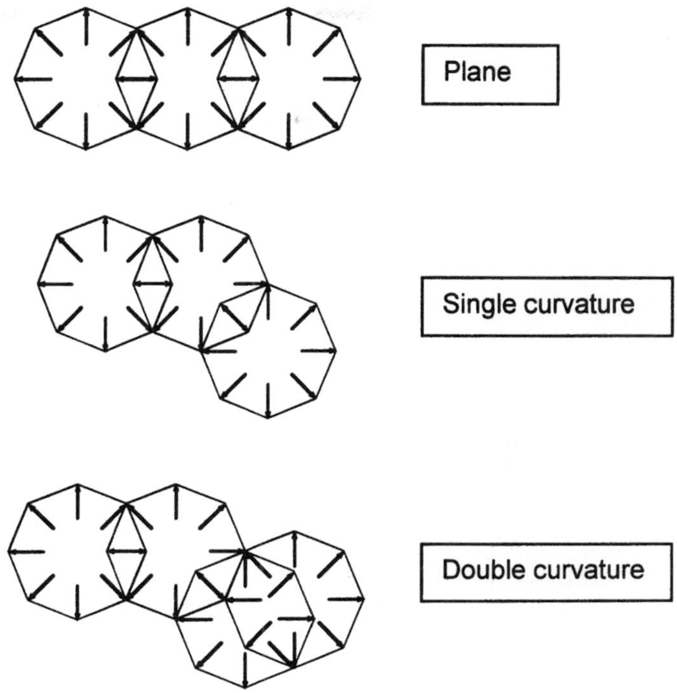

Figure 2.17 *Structural situation diagrams*

When transposed in three-dimensional space, these diagrams open many morphologic possibilities bringing together properties of pneumatic systems and properties of tensegrity systems.

2-233.2 Double layer grids

Many controversies arose on double layer grids since in many cases struts were separately considered as components and touch each other at a common node. This was not the case for a proposal by Hanaor (Figure 2.18), which was built on a node-on-cable principle. But it was also true that in this configuration the mechanical behaviour of the grid under external actions was not satisfactory.

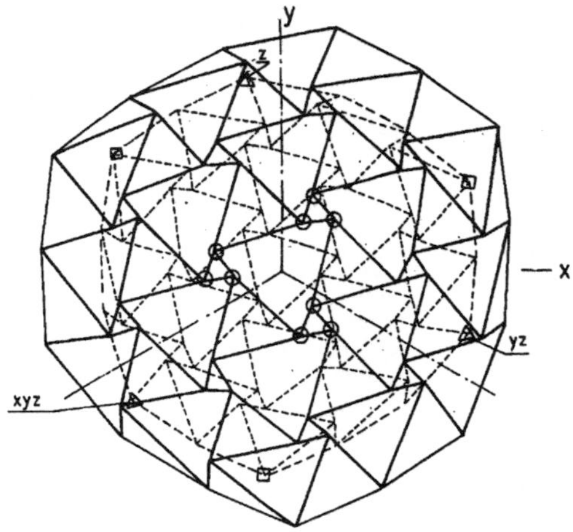

Figure 2.18 *Hanaor double layer grid (dome)*

Recently, Wang [Ref 2-18] suggested using the expressions "non-contiguous" or "contiguous" tensegrity systems. This was interesting but not sufficient since these expressions pre-supposed that a chain of compressed struts can not be considered as a compressed component.

In order to illustrate our definition we have taken as an example the double layer grid that we designed some years ago (Figure 2.19). According to our definition the following decomposition can be made. The whole system comprises a continuum of cables and five compressed components, each is made with a set of struts, and these five sets do not touch each other. There are two kinds of compressed components: one with four struts linked at one node and four with three struts linked as a chain.

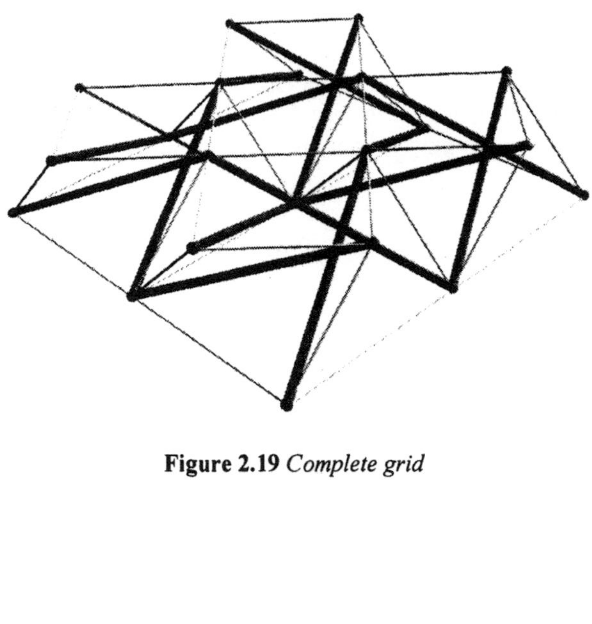

Figure 2.19 *Complete grid*

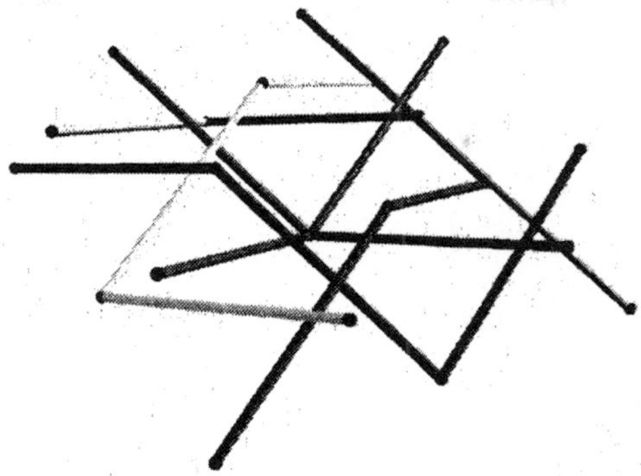

Figure 2.20 *Five compressed components*

This kind of grid can be included in the definition of double layer tensegrity grids. And this is also the case for many double layer grids in which discontinuous compressed components can be identified.

2-233.3 Cable domes

Cable domes have been developed in recent years. The names of Geiger and Levy are associated with these structures. It is clear that they have been inspired by the tensegrity principle.

But at the end of the day the result seems to be in the scope of pre-stressed systems. The whole cable-strut net is associated with a huge compression ring, just as membranes are tensioned on fixed masts or beams. They can not be strictly classified as tensegrity systems. Two kinds of compressed components can be identified: vertical struts and compressed ring (Figure 2.21). This last component is on the boundary of the system and not inside it, which excludes these systems from the tensegrity classification (if you agree with our "extended" definition). And this is not only a matter of agreement since intrinsic properties of these two classes are not the same.

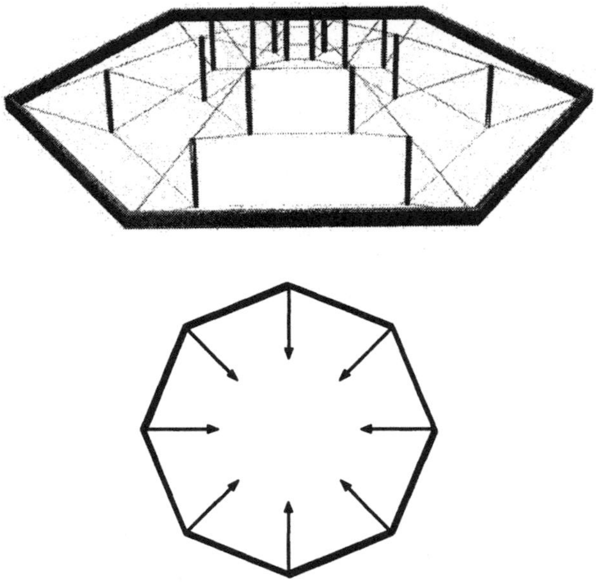

Figure 2.21 *Cable dome principle*

Generally the compressed ring is made of reinforced concrete, sometimes in pre-stressed concrete and its size is not comparable with other components (Figure 2.22). Moreover, these rings can be a part of the whole building, and it can be difficult to identify them as a separate entity. But so far as structural performance is concerned, it is obvious that these cable-domes are very efficient.

Figure 2.22 *Cable domes with four cable hoops*

2-233.4 Recent proposals

In recent years some new proposals have been submitted and described as tensegrity systems. In Switzerland, Passera and Pedretti built some experimental systems [Ref 2-19] and developed calculations on several tensegrity systems in order to participate in Expo 02. One of these structures is represented in Figures 2.24 and 2.25.

These systems are based on an elementary module consisting of five struts and height cables (Figure 2.23). Four of the five struts describe a quadrangular polygon. This elementary octahedral cell is not a tensegrity system, according to the submitted definition. Compressed components lie on the boundary of the system. Moreover, there exists in these cases mechanisms of higher order.

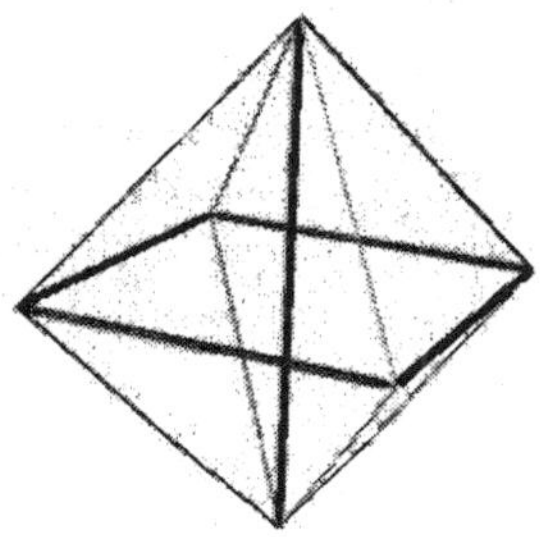

Figure 2.23 *Octahedral cell*

Nevertheless, if we look at some examples of assemblies it is still sometimes possible to satisfy the "extended" definition. In the case of systems evoked by Passera and Pedretti, two kinds of compressed components can be identified: several posts (quoted "A"), which do not touch each other and a larger compressed component, with some elements like "B", which are lying on the edge. It is certainly possible by mechanical inspection to find one (or more) self-stress-states such that all of the elements on the boundary are under tension. By removing hollow profiles like "B" and replacing them by cables (or at less components with unilateral rigidity in tension), the whole system could be qualified as a "tensegrity system".

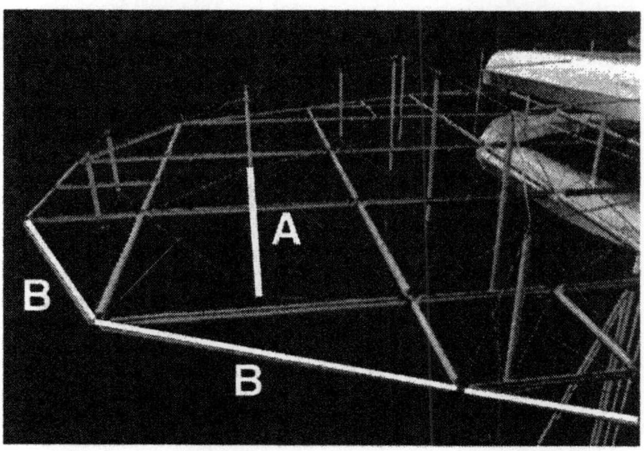

Figure 2.24 *Passera–Pedretti structure: perspective view*

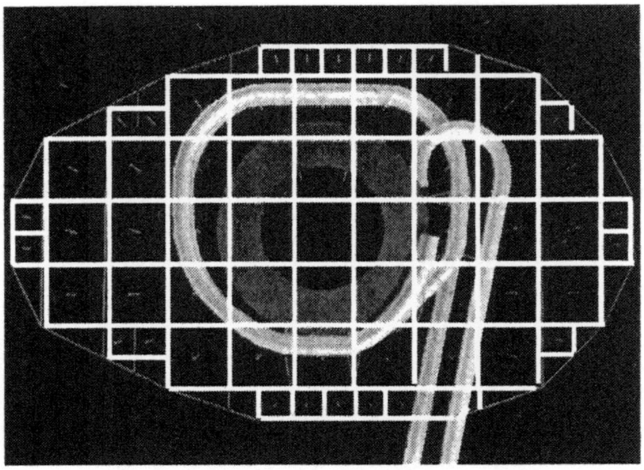

Figure 2.25 *Passera–Pedretti structure: plan view*

2-233.5 Endothelial cells

The last example is not in the scope of construction architecture, but in the biological field. Endothelial cells comprise the kernel, the cytoplasm and the surrounding membrane. The cytoskeleton is composed of microtubules and actin filaments among other components. Recently, Donald E. Ingber [Ref 2-20] developed his theory on the mechanotransduction through the cytoskeleton, suggesting an analogy between the mechanical behaviour of cytoskeleton and tensegrity systems' mechanical behaviour. It appears that we can indeed find a limited degree of common ground. The composition of the cytoskeleton allows us to speak of a tensegrity system in this case. Moreover, the whole endothelial cell can be considered as a tensegrity system, which satisfies the definition: the kernel is also a compressed component and the membrane the boundary tensioned surface. It is not surprising that Ingber's proposal for modelling this cell be a six-strut tensegrity system including another one of lesser size [Ref 2-21]. But this kind of analogy must be carefully verified.

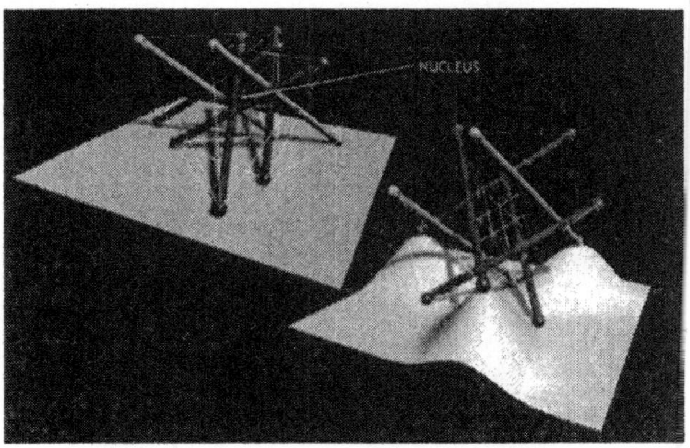

Figure 2.26 *Cell modelling with tensegrity systems*

2-3. Conclusion

It is not an easy task to define Tensegrity Systems. It could be claimed, as Fuller did, that everything in the Universe is tensegrity, but this would be confusing: no distinction could be made, for instance with systems working on the basis of compression. Tensegrity systems have intrinsic properties related to the continuum of tensioned components, and they have to be enhanced. The extended definition was necessary since it was obvious that many systems were excluded on the basis of the early patents. We will omit the qualification "extended" in the following chapters and we will refer to "patent" definition in those cases which require this restricted definition.

References

Ref 2-1. Emmerich D.G., Structures tendues et autotendantes – Monographies de Géométrie Constructive, Éditions de l'Ecole d'Architecture de Paris La Villette, 1988.

Ref 2-2. Lodder C., The Transition to Constructivism. The Great Utopia. The Russian and Soviet Avant-Garde, 1915–1932, exhibition, Guggenheim Museum, 1992.

Ref 2-3. Bouderbala M., Systèmes à barres et câbles pliables, dépliables, Diplôme d'Etudes Approfondies, Université Montpellier II, 1994.

Ref 2-4. Motro R., "Tensegrity Systems: State of Art", *International Journal of Space Structures* (Special Issue on Tensegrity Systems), R. Motro Guest Editor, Vol. 7, N° 2, 1992.

Ref 2-5. "Morphology", *International Journal of Space Structures* (Special Issue on Morphology), H. Lalvani Guest Editor, Vol. 11, N° 1 & 2, 1996.

Ref 2-6. Maculet R., Etude et calcul de structures autotendantes, Diplôme d'Architecture, École d'Architecture de Paris La Villette, 1987.

Ref 2-7. Fuller R.B., Inventions: The Patented works of R.B. Fuller, St. Martin's Press, 1985.

Ref 2-8. Chu R., "Tensegrity", *Journal of Synergetics*,Vol. 2, N° 1, June 1988.

Ref 2-9. Emmerich D.G., "Charpentes Perles" ("Pearl Frameworks"), Institut National de la Propriété Industrielle (Registration n° 59423), 26 May 1959.

Ref 2-10. Snelson K., "Continuous tension, discontinuous compression structures", U.S. Patent No 3,169,611, February 16, 1965.

Ref 2-11. Fuller R.B., Synergetics explorations in the geometry of thinking, Collier Macmillan Publishers, London, 1975.

Ref 2-12. Edmondson A., "Geodesic Reports: The Deresonated Tensegrity Dome", Synergetica, *Journal of Synergetics*, Vol. 1, N° 4, November 1986.

Ref 2-13. Pugh A., An introduction to tensegrity, University of California Press, Berkeley, 1976.

Ref 2-14. Tardiveau J., Siestrunck, R., "Efforts et déformations dans les assemblages en treillis critiques et surcritiques en élasticité linéaire", Note présentée par M. M. Roy à l'Académie des Sciences, T. 280, N° 3, 20 January 1975.

Ref 2-15. Roth B., Whiteley W., "Tensegrity Frameworks", Transactions of the American Mathematical Society, Vol. 256, N° 2, 1981, pp. 419–446.

Ref 2-16. Motro R., Formes et Forces dans les Systèmes Constructifs, Cas des Systèmes Réticulés Spatiaux Autocontraints, Thèse d'Etat, Université Montpellier II, June 1983.

Ref 2-17. Vassart N., Laporte R., Motro R., "Determination of mechanisms's order for kinematically and statically indeterminate systems", *International Journal of Solids and Structures*, Vol. 37, 2000, pp. 3807–3839.

Ref 2-18. Wang B.B., Yan Yun Li, "Definition of tensegrity systems. Can dispute be settled?", Proceedings of LSA98 "Lightweight structures in architecture engineering and construction", edited by Richard Hough & Robert Melchers, ISBN 0 9586065 0 1, Vol. 2, 1998, pp. 713–719.

Ref 2-19. Pedretti M., "Smart Tensegrity Structures for the Swiss Expo 2001", Proceedings of LSA98 "Lightweight structures in architecture engineering and construction", edited by Richard Hough & Robert Melchers, ISBN 0 9586065 0 1, Vol. 2, 1998, pp. 684–691.

Ref 2-20. Ingber D.E., "Tensegrity: The Architectural Basis of Cellular Mechanotransduction", *Annu. Rev. Physiol.*, 1997, pp. 575–599.

Ref 2-21. Ingber D.E., "The Architecture of Life", *Scientific American*, January 1998, pp. 30–39.

3

Fundamental Concepts

3-1. Introduction

The question of tensegrity stability is certainly the first that crops up for many people, who are "surprised", when they discover these systems for the first time. This was my the first question which arose for me when I began, but I was convinced that there was not any particular "secret" to be uncovered – rather only some fundamental concepts to rediscover. The aim of this chapter is to provide the "keys", which unlock the door to the understanding of tensegrity.

Aesthetic and mechanical characteristics of tensegrity systems result from the continuity of the tensioned components set and from the discontinuity of compressed components: the concept of relational structure and the associated use of the graph theory clearly qualify these continuity and discontinuity concepts.

The stiffness of tensegrity systems is conditioned by the stabilisation of infinitesimal mechanisms with states of self-stress. This stiffening is only possible with geometry which is consistent with static equilibrium criteria. This is why it is useful to explain the meaning of mechanisms (finite and infinitesimal mechanisms) and of associated states of self-stress. The stabilisation of infinitesimal mechanisms plays an important role since it also explains why compressed and tensioned components cannot be exchanged by simple duality.

These fundamental concepts are developed in the following paragraphs. They are illustrated in the example of "elementary equilibrium"[1].

3-2. Relational structure

All reticulate systems being created by an assembly of components, it is necessary to define this mode of assembly, particularly when these components are themselves comprised of several sub components.

---

[1] When there is no risk of confusion we do not differentiate between elements and components in this chapter.

It was suggested in the "patent" definition (Chapter 2) that we should take into account the straight segments defined by the ends of compressed or tensioned elementary components (like struts and cables). Allocating to each segment "b" a place between two nodes "i" and "j" fulfils this description. This description, requiring no dimensional geometrical datum, is called "relational structure", [Ref 3-1].

The necessarily even number of nodes characterises the case of tensegrity systems with struts. If "n" is this number, and if attention is paid to the spatial case, the minimal value of n is 6 (the case n = 2 corresponds to a linear system, the case n= 4 to a bi dimensional system). Graph theory can be useful for defining the relational structure of this elementary system, which is described by the wording of the nodes and their links. The choice among multiple relational structures that can be constructed with six nodes is conditioned by the nature of tensegrity systems, since tensioned elements constitute a continuous set and compressed elements a discontinuous one. This leads to a six nodes complete plane graph from which the graph of tensioned elements will be extracted[2]. This graph comprises four links at each node (Figure 3.1).

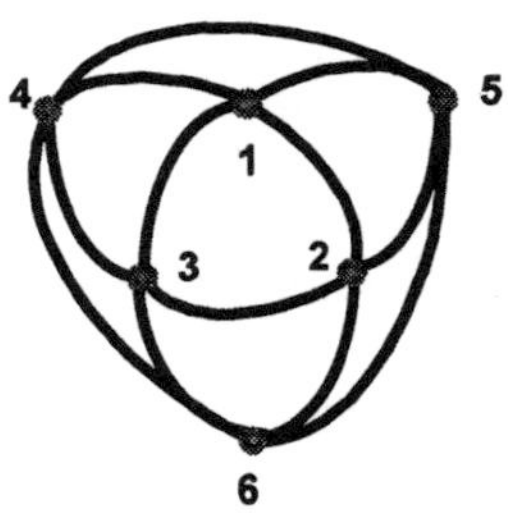

Figure 3.1 *Six nodes complete plane graph*

But, qualitatively speaking, three links are sufficient to ensure the necessary condition of spatial stability for a node which, furthermore, will receive one (and only one) compressed element[3]. This last element will be placed inside a solid angle defined by the three tensioned links, which must not be coplanar to preserve spatiality (Figure 3.2).

[2] The complete set of tensioned elements can be applied on a sphere; on the basis of this homeomorphy. It can be demonstrated that the associated graph is necessarily plane. The given graph is complete, in terms of a plane graph. No more links can be added. The complete graph with six nodes would not be plane, since some links would intersect each other.

[3] It is possible to have only two cables for specific cases of some systems ("star systems" or "stella octangula"), described in the "Topologies" chapter.

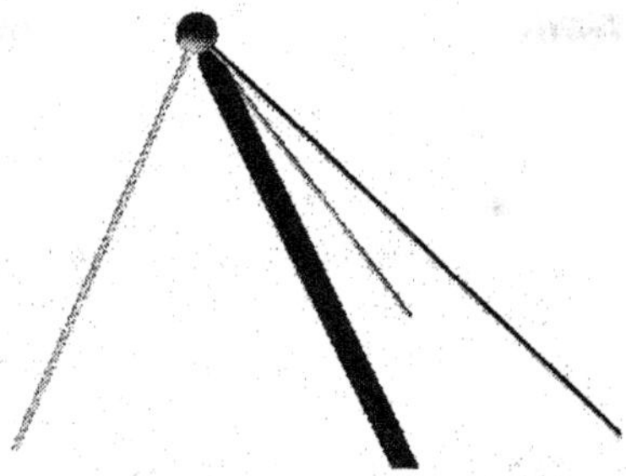

Figure 3.2 *Necessary condition of spatial stability for a node*

This remark allows us to eliminate a link by node in the six nodes of the complete plane graph, and thus to obtain the relational structure of elementary equilibrium by superposition of this continuous plane graph (tensioned elements) on the disjoint graph of the three compressed elements (Figure 3.3).

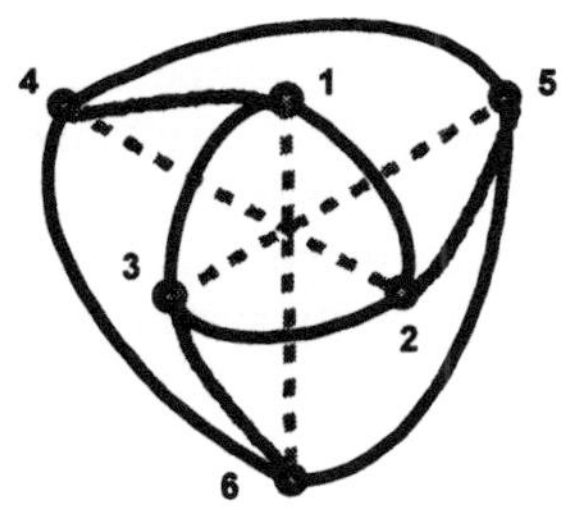

Figure 3.3 *Elementary equilibrium: total graph of the relational structure*

In Table 3.1, we describe this relational structure that comprises three disjoint compressed elements and nine tensioned elements.

Table 3.1 *Elementary equilibrium: relational structure*

Links	Nodes	
Cables		
1	1	2
2	2	3
3	1	3
4	1	4
5	2	5
6	3	6
7	4	5
8	5	6
9	4	6
Struts		
10	2	4
11	3	5
12	1	6

3-3. Geometry and stability

3-3.1. Introduction

The geometry of a spatial reticulate system is completely defined by its relational structure and by the knowledge of the coordinates x, y, and z for its "n" nodes in reference to a chosen axis system.

The stability of tensegrity systems can be satisfied only for geometry in which a situation of stable static self-equilibrium can be established: the study of tensegrity systems necessitates a "form-finding process" which allows us to attain such geometric equilibrium.

3-3.2. The balloon analogy

Analogical reasoning can help us to simply explain the functioning of a tensegrity system and the relationship between geometry and stability. A balloon is a familiar object for all of us; it consists of an envelope whose shape is determined by its manufacture (it results from an assembly of membrane plane pieces or "strips").This shape is able to enclose a volume of air, which is equal to that of the envelope, once the balloon is inflated and the envelope deployed (Figure 3.4). Furthermore, a balloon can be considered as a tensegrity system since it is a stable self-balancing system made up of two components: a compressed component, the air and a tensioned component, the membrane.

Figure 3.4 *Football and rugby balloons*

To study various geometric situations and associated shapes, several cases should be considered[4]:

A The balloon has no "form" – nor is its geometry known, when the blown air volume remains less than the volume that the envelope can enclose. By pushing here or there, one can obtain different shapes (Figure 3.5). The form is called *indeterminate form*.

[4] Self-weight is neglected in this explanation.

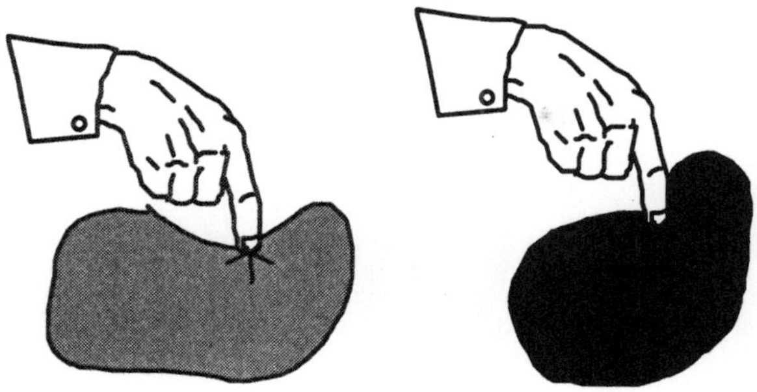

Figure 3.5 *Indeterminate shape of a non-inflated membrane*

B The balloon takes its shape as soon as the blown air volume becomes equal to the volume which can be enclosed in the envelope[5]. The corresponding geometry is an equilibrium geometry (Figure 3.6); air pressure is identical inside and outside the membrane. In this second state the balloon again remains inappropriate for its subsequent use, it has no stiffness.

Figure 3.6 *Equilibrium geometry*

C It is by inflating again, that is to say by blowing a volume of air greater than that of the envelope (and therefore by increasing the internal pressure) that a rigid object is obtained. In this third state the air is under pressure and the membrane that constitutes the envelope is under a state of tension. The pressure of the air confers to the balloon a "stiffness", which is a function of its level. The existing tension in the membrane slightly deforms it, but the envelope holds its initial shape (unless, of course, a too-high-pressure had made it burst!) (Figure 3.7). There is a balance between the air pressure and the membrane tension.

---

[5] Geometrically speaking, this shape is related to the manufacture of the envelope: it is different, for instance, for a football or a rugby ball. More generally speaking, each inflatable membrane will provide a specific shape, like those inflatable cushions used in aeroplanes.

Figure 3.7 *Inflated balloon and possible collapse*

The analogy between the different shapes that have been described for a balloon (and which are valid for any inflated membrane), and a tensegrity system can be established by associating with the membrane a system of "envelope" element net that could be tensioned, and by replacing the included air by internal elements susceptible to be compressed.

It is possible, and useful, to create a physical model that allows us to follow the development of the system. For such a model, it is necessary for all internal elements to have the same length "s" and for the external elements all to have the same length "c". One can thus constitute compressed elements with variable length (telescopic element) by associating two aluminium tubes with different diameters and by creating a screw-tightening device (Figure 3.8). The length of the compressed elements has to be between one and two times the length of the tensioned elements.

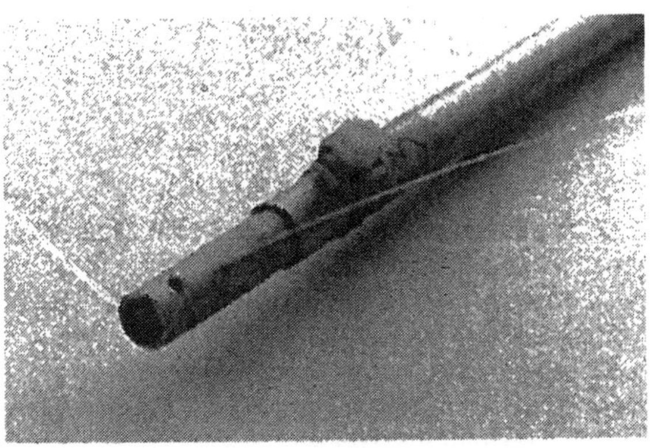

Figure 3.8 *Telescopic element*

The holes at the two ends of the telescopic elements receive nylon fishing lines with a stopping device (a simple knot suffices).

In this physical model (and by analogy with the balloon) the telescopic elements play the role of the air and the fishing lines the role of the envelope. One thus finds the three states that have been identified for the balloon.

A If the length of the internal elements (those playing the role of air) is insufficient (leading for example to a ratio "r" of 1.414 between the external and internal elements) the set of envelope elements will not have a definite shape; it will be possible by an external action to obtain several shapes. In terms of mechanics the system is then called "cinematically indeterminate". A first geometry is defined as a triangular prism (Figure 3.9). This system is unstable, and another shape can be defined (Figure 3.10). Geometrically speaking, the six nodes are apices of an octahedron. In this last case, for such geometry, the three diagonals of the octahedron intersect each other at the same point. The way from one shape to another results from the existence of a "finite mechanism", that can, for example, be activated by a relative rotation of the two triangles of tensioned elements lying in parallel planes. The first and the second geometry preserve the same lengths for elements, but they are different.

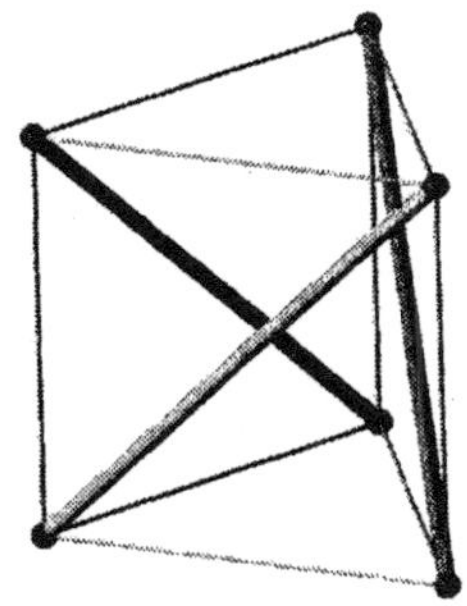

Figure 3.9 *Cinematically indeterminate system (finite mechanism). First geometry*

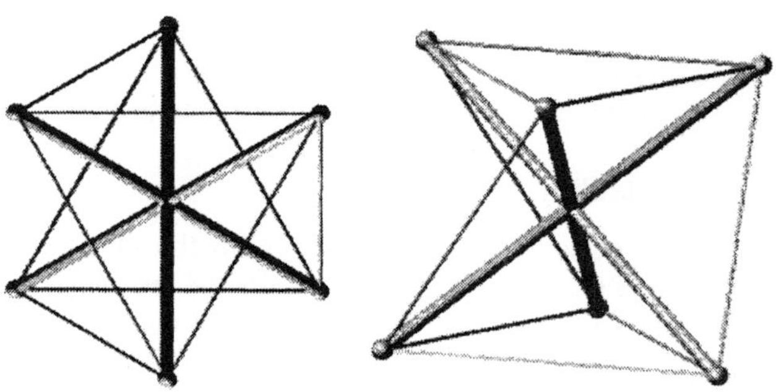

Figure 3.10 *Cinematically indeterminate system (finite mechanism). Second geometry*

B For a given value of the strut length (corresponding precisely to r = 1.468), the totality of the cable net takes a singularly definite shape, which will be referred to as "null self-stress equilibrium geometry". This geometry, in the general case, can be calculated by a process called "form-finding". In this state, no constitutive element is stressed: the geometric distance between the nodes corresponds strictly to the length of manufactured elements, as was the case previously. The difference is, however, that at this time only one feasible geometry exists (from a mechanical point of view the system is again cinematically indeterminate, but this time the mechanism is infinitesimal). The geometry of the elementary equilibrium is a "null self-stress equilibrium geometry"[6] (Figure 3.11).

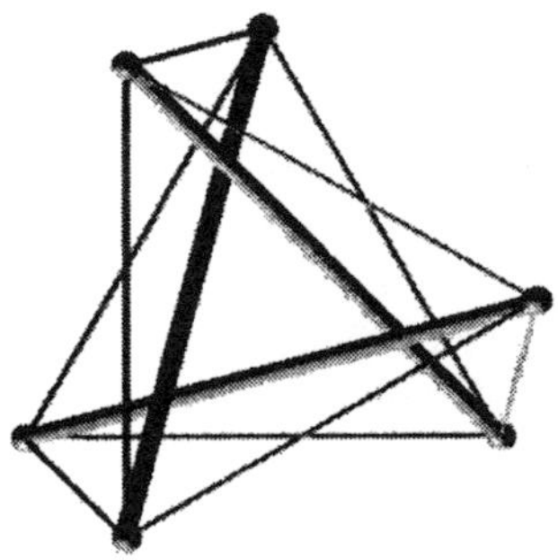

Figure 3.11 *Null self-stress equilibrium geometry*

C The third phase corresponds to an increase of length for the internal elements: these elements are then compressed and elements on the envelope are simultaneously tensioned. One can observe on the physical model that it becomes difficult to increase this length. The corresponding geometry will be deformed when compared with "null self-stress equilibrium geometry" but will keep its overall characteristics (unless there is a collapse of tensioned elements on the envelope or a buckling of the internal element!). The system is then self-stressed: the self-stress qualifies the load effects of compression in the internal elements and the load effect of tension in envelope elements. Its level conditions the stiffness of the tensegrity system. In fact the denomination "tensegrity" qualifies only the tensional integrity[7]. This last geometry will be referred to as "geometry deformed by self-stress", in fact it is rather similar to the previous one.

3-3.3. Geometry and stability

In fact, in this simple example the analogy rests on the fact that the telescopic element plays the role of the air under pressure while the tensioned elements can be associated with the balloon's envelope. Two geometrical ranges can be identified. If one plots on a graph the relationship between the ratio "r" of element lengths s/c; and the relative rotation θ between the inferior and superior triangles of the

---

[6] Relative rotation between the two triangles is equal to 30°.

[7] This denomination does not reflect the existence of compressed elements. The French word "tenségrité" is the translation of the word "tensegrity" invented by R. B. Fuller.

elementary equilibrium, the straight horizontal line corresponding to the value 1.468 separates a zone where there exists only one possible geometry, from another zone where two geometries correspond to a single value of the ratio (Figure 3.12).

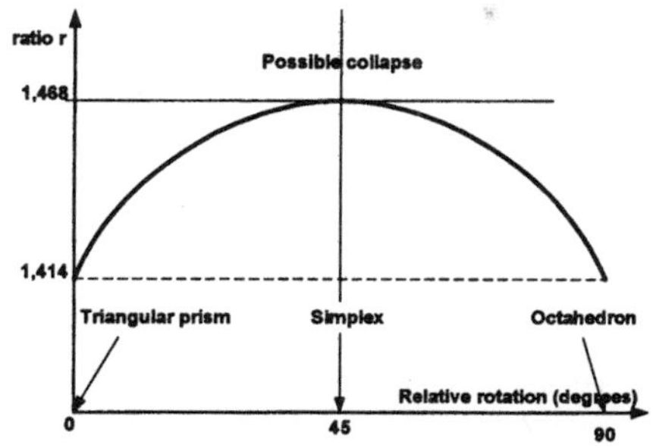

Figure 3.12 *Relationship between the ratio "r" and the rotation θ*

This result can be verified on the physical model.

The limit value of the ratio, 1.468 in this case, corresponds to the "null self-stress equilibrium geometry": this ratio value results from a form-finding process[8].

Geometries corresponding to lower values are not defined in a single way: finite mechanisms exist that allow a path from one geometry to the other. A length ratio value equal to $\sqrt{2}$, as can be verified (for example with the physical model), gives access to two feasible geometries.

Geometries corresponding to higher values of the length ratio (more than 1.468) are defined in a single way: the whole system is self-stressed and the elements deformed. These geometries are, in fact, very close to the "null self-stress equilibrium geometry".

It is useful to note on the physical model that in this last configuration (ratio over 1,468), there again exists small possible movements that one can activate by acting on one or more nodes: these movements correspond to infinitesimal mechanisms. But once the action is stopped, the system re-establishes its initial geometry, thanks to the self-stress that stabilises such infinitesimal mechanisms.

3-3.4. Form-finding

The preceding paragraphs have tended to underline the importance of the "null self-stress equilibrium geometry", whose knowledge results from a process called "form-finding". In some simple cases one can reach this geometry by researching the maximum of the length of the compressed elements compatible with that of the tensioned elements; this "cinematic" method is detailed in Chapter 5. It is equally

[8] See the demonstration in Chapter 5.

possible to associate a geometrical condition with an equilibrium condition for some tensegrity systems such as the elementary equilibrium.

If one considers, for example, one node of the system such as node 4, it is necessary to ensure its equilibrium. This node (Figure 3.13) is submitted to actions exerted according to directions defined by links, that we have noted 4, 7, 9 and 10 in Table 3.1, which describes the relational structure.

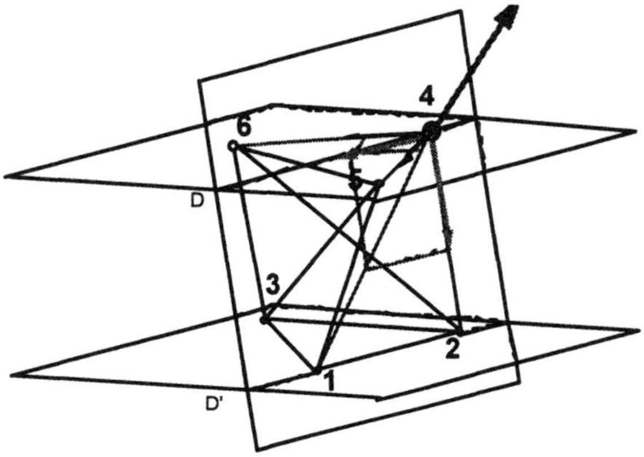

Links	Nodes	
Cables		
1	1	2
2	2	3
3	1	3
4	1	4
5	2	5
6	3	6
7	4	5
8	5	6
9	4	6
Struts		
10	2	4
11	3	5
12	1	6

Figure 3.13 *Equilibrium of node 4*

By reason of symmetry in this system, links 7 and 9 can be replaced by an equivalent virtual link, noted 13 and supported by the bisecting line D of the angle 546. In order to ensure the equilibrium of node 4, the supports of the three actions, noted 4, 10 and 13, have to be coplanar[9]. Plane P, defined by two of them, 4 and 10, intersects the lower triangle plane according to the line D' between nodes 1 and 2 and the upper triangle one according to the bisecting line D. The two planes containing these triangles are parallel and consequently lines D' and D are also parallel (this parallelism is preserved in horizontal projection), and this sets the relative rotation of the two triangles to 30°[10].

[9] In order to ensure the equilibrium of a material point subject to three actions, their supports have to be coplanar.

[10] This result is of course identical to that given by the "kinematic" method. It is nevertheless necessary to note that it does not condition the distance between plans of upper and lower triangles; the elementary equilibrium represents the particular case for which this distance corresponds to identical lengths of all tensioned elements. This result can be extended to the case of tensegrity systems defined by two polygons of n edges situated in two parallel plans and linked simultaneously by tensioned external elements and internal compressed elements.

3-4. Self-stress states and mechanisms

3-4.1. Introduction

Learning from the preceding approach, it appears that two mechanical concepts dispense with questions relating to the stability of tensegrity systems; namely self-stress and mechanism concepts that are clarified in the following paragraphs. The study of their relationship is also developed.

3-4.2. Cinematic indeterminacy

When several elements are assembled, the choice of manufacture lengths and that of relational structure can lead to a cinematic indeterminacy: the geometry of the assembled system is not defined in a single way. The path between the different geometries follows a movement of elements; this movement, which requires no energy (at first order), takes the name "mechanism". If, in the course of the movement there is no length variation for the elements, then the mechanism is called "finite". But if, on the other hand, length variations are observed (even very small) then the movement corresponds to an "infinitesimal" mechanism. These two notions that are useful for the study and the understanding of tensegrity systems are illustrated with simple examples. These distances and lengths are measured by assuming that nodes are points on which punctual ends of elements are superposed. This hypothesis allows us to forget the real dimensions of the nodes.

3-42.1. Finite mechanism

Let us consider a set of four elements with compression rigidity; their assembly with four hinges constitute the "envelope".

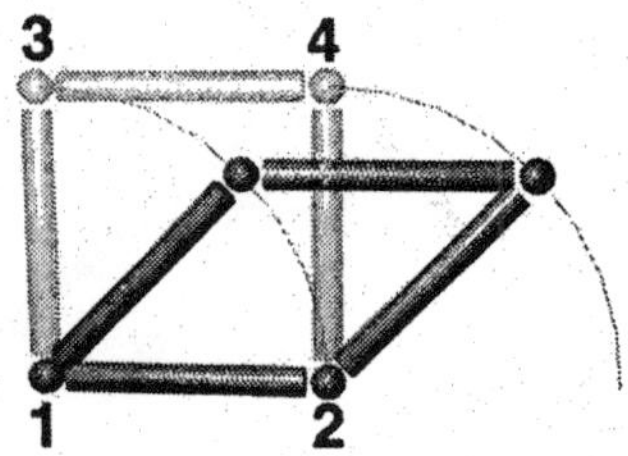

Figure 3.14 *Finite mechanism*

So as to eliminate the effects of self-weight these four elements are put on a horizontal plane. The two extremity nodes of a same element are supposed to be completely fixed (1 and 2 for example), and this avoids any overall displacement.

A system such as this is cinematically indeterminate: several geometries corresponding to circular trajectories of nodes 3 and 4 can exist (Figure 3.14).

3-42.2. Infinitesimal mechanism

Let us consider two rectilinear elements assembled end to end by one common extremity, the other being fixed; it is possible to transversally displace the common node of a distance d, with infinitesimal absolute deformations of elements (they are in fact proportional to the square of d). This movement is known as "infinitesimal mechanism" [Ref 3-2].

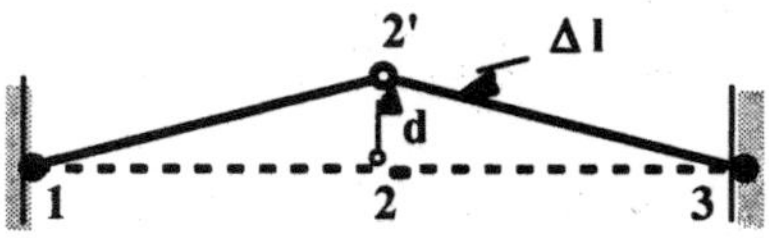

Figure 3.15 *Infinitesimal mechanism*

3-4.3. The static indeterminacy: pre-stress and self-stress

3-43.1. Static indeterminacy

Let us consider a set of four elements with a common node. The other end of each element is assumed to be fixed[11] (Figure 3.16). Values d_{ij} are the geometric distance to be reached at the end. One can imagine, in the first instance, that manufacture lengths l_{ij} satisfy the relationship:

E 3-1 $$l_{ij} = d_{ij},$$

The three equations of static are generally insufficient to determine load effects of elements when an action is applied on the node: the system is statically indeterminate (hyperstatic)[12].

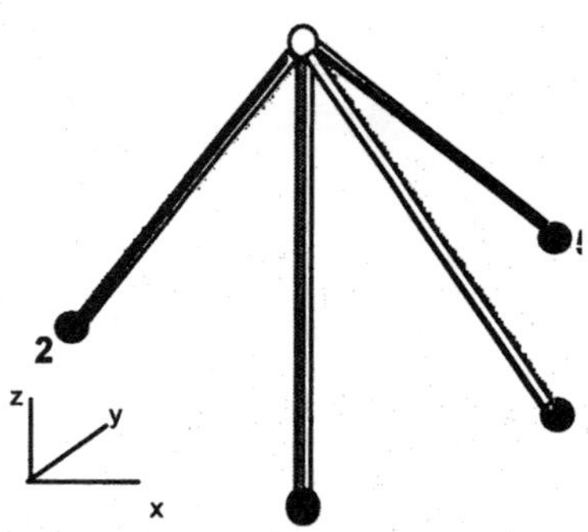

Figure 3.16 *Statically indeterminate system*

Before any application of external action, it is possible to envisage the case where the length of one of the elements would be different from the distance between

[11] Conventions for nodes representation and stresses are developed in Appendix A.
[12] Taking into account element strains allows us to solve the problem.

nodes to which one wishes to link it (Figure 3.17). The assembly created in such conditions induces an initial constraint state or "pre-stress". The insertion of element "14", whose manufactured length l_{14} is less than distance d_{14}, is equivalent to the application of two opposite forces: one on node "1", the other on node "4". The last one has no effect on the system. The first one introduces stresses in the three other elements thus creating an initial state of stress: a pre-stress. The pre-stress effect will be combined with the effect of the actions applied afterwards on the common node "1". The implementation of this pre-stress cannot be established without fixing nodes 2, 3, 4 and 5.

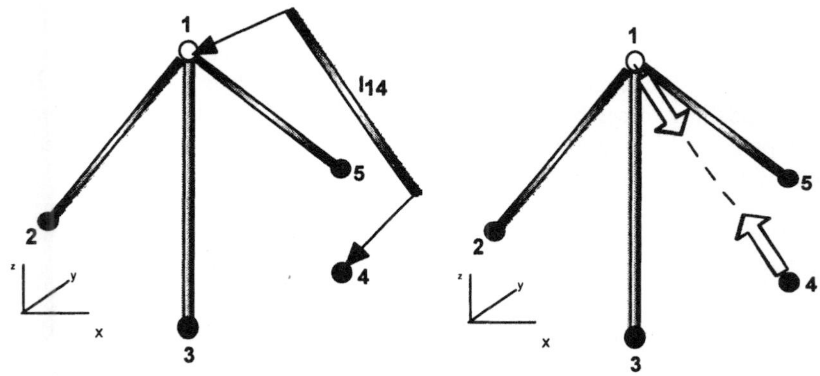

Figure 3.17 *Initially stressed state*

One can envisage another example, for which such an initially stressed state no longer depends on conditions of fixing be they whole or part of the nodes: the corresponding system is known as "self-stressed"[13].

3-43.2. Self-stressed systems

In the definition of tensegrity systems, we specified that they are in "state of stable self-equilibrium". This state describes the equilibrium of nodes. This self-stress is tridimensional, since most tensegrity systems are spatial systems. Linear and plane systems are just specific cases. Among the three chosen examples developed below we illustrate the concept of self-stress, from linear self-stress to tridimensional self-stress.

3-432.1 Linear self-stress

The simplest case of self-stress is the linear self-stress that is illustrated by the example comprising a tensioned element, cable, and a compressed element, strut (Figure 3.18).

Figure 3.18 *Linear self-stress*

---

[13] A global displacement can be operated, but it has no effect on the initial stressed state.

In this case, the compressed element is a circular cross-section tube; nodes can be metallic discs placed at the extremities of the tube, the cable being placed inside.

Three cases are to be considered according the relative length values "c" for the cable and "s" for the tube, which are manufacture lengths of these elements.

- First case c > s, the cable is not in tension, its shape is not defined.
- Second case c = s, the cable is not in tension. Its shape is rectilinear; the geometry of the system is defined. This geometry is the "null self-stress equilibrium geometry".
- Third case c < s, the cable and the tube are submitted respectively to a simple load effect of tension and compression. The equilibrium of the whole system necessitates the equality between the absolute values of the two internal corresponding forces N_s and N_c. The geometry of the system is deformed when compared with equilibrium geometry. The system is self-stressed. The equilibrium for each element is illustrated in Figure 3.19.

One can observe in this first example that the value of the ratio r= s/c characterises the system.

- When r< 1, the shape is not defined (cinematically indeterminate system).
- When r= 1, the shape is defined (null self-stress equilibrium geometry).
- When r> 1, the shape is defined, the system is in state of self-stress.

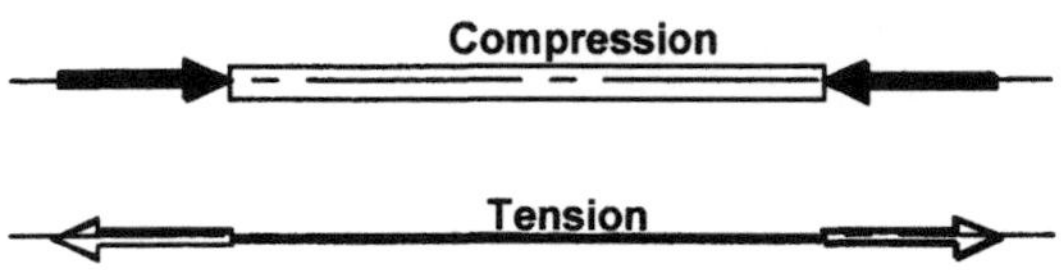

Figure 3.19 *Equilibrium of the two components*

3-432.2 Bidimensional self-stress

Bidimensional self-stress can be illustrated with a system comprising four struts and two cables. The equilibrium geometry is that of Figure 3.20 (struts have identical length "s", and cables identical length "c"). The three cases studied for the preceding example can be summed up by again noting that r= s/c.

- For r< 1.414, the shape is not defined (cinematically indeterminate system).
- For r= 1.414, the shape is defined (null self-stress equilibrium geometry).
- For r> 1.414, the shape is defined; the system is in a non-null self-stress-state. The equilibrium self-stress-state corresponds to the diagram in Figure 3.20.

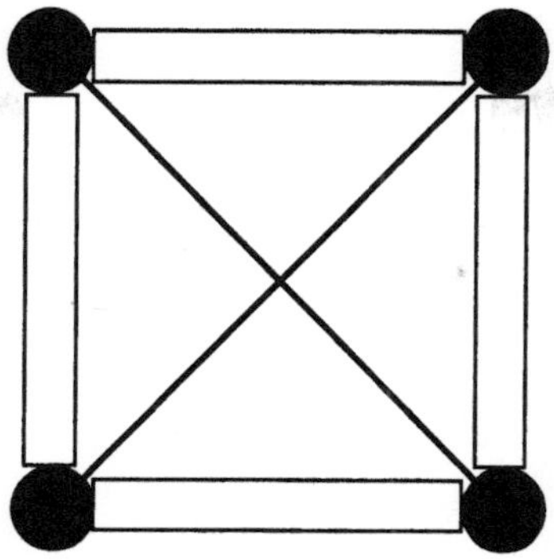

Figure 3.20 *Equilibrium of bidimensional self-stress*

Study of equilibrium of each node leads to relationship[14]:

E 3-2 $N_c = 1{,}414\ N_s$

This is deduced, for example, from the dynamic diagram of actions applied on node (Figure 3.21).

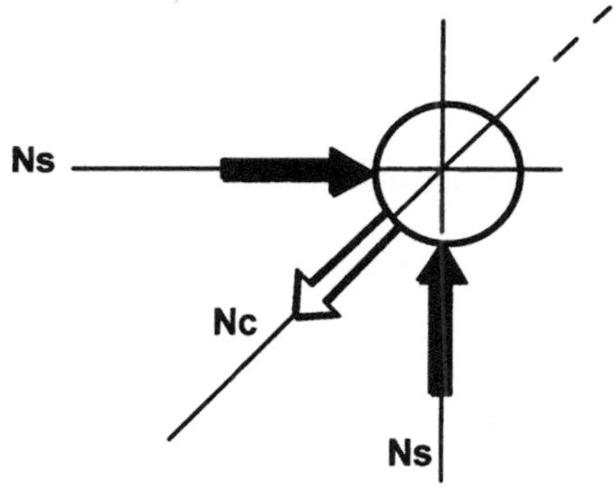

Figure 3.21 *Equilibrium of a node*

3-432.3 Tridimensional self-stress

One can extend the notion of self-stress to tridimensional space by taking as an example a configuration constituted by an octahedron whose edges are the tensioned elements, and diagonals the compressed elements (Figure 3.22).

[14] This equation and the associated figures are defined with absolute values: action on node and value of internal forces in components (struts or cables) have the same absolute value.

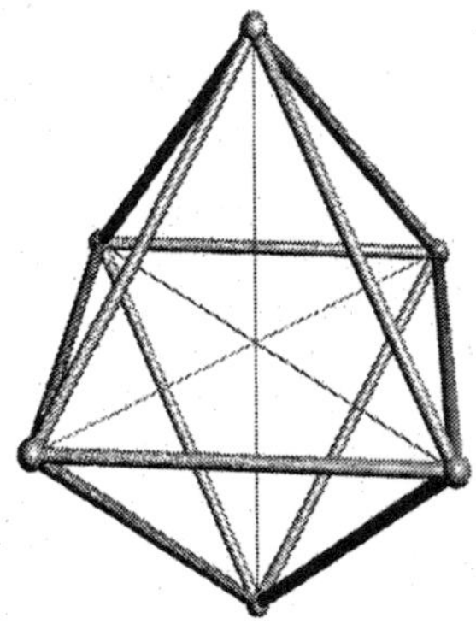

Figure 3.22 *Tridimensional self-stress*

The octahedron is rigid by itself; the corresponding geometry is what we have referred to as "equilibrium geometry". The ratio "r" can be simply calculated in this case; it is equal as in the preceding case to 1.414. If it is of greater value, the system will be self-stressed.

This example is not an example of a tensegrity system, but it has been chosen to illustrate tridimensional self-stress, because of the clarity of its static equilibrium diagram, even if it necessitates a non-realistic crossing of diagonals. Tensegrity systems are a subclass of tridimensional systems in a state of self-stress.

3-4.4. Infinitesimal mechanism stabilisation

The last concept, which is necessary to help in our understanding of the behaviour of tensegrity systems, is the concept of infinitesimal mechanism stabilisation. If we examine the example of the two elements assembled end to end, it is possible to try to stabilise the infinitesimal mechanism by introducing a pre-stress.

If the two elements are chosen in such a manner that the subsequent relation is satisfied:

E 3-3 $\quad l_{ij} < d_{ij}$

a state of tension pre-stress is established (Figure 3.23).

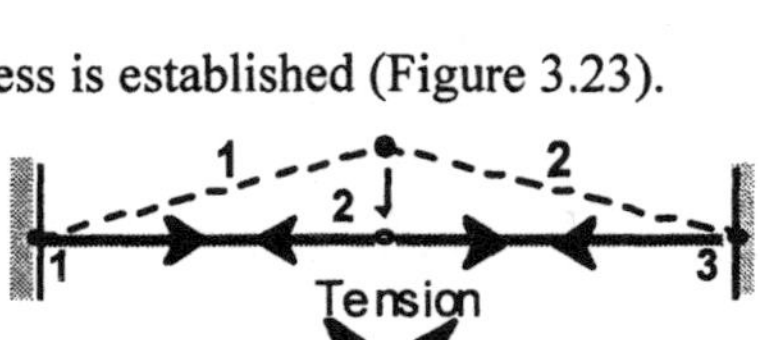

Figure 3.23 *Tension pre-stress*

If the two elements are chosen in such a manner that the subsequent relation is satisfied:

E 3-4 $\quad l_{ij} > d_{ij}$

a state of compression pre-stress is established (Figure 3.24).

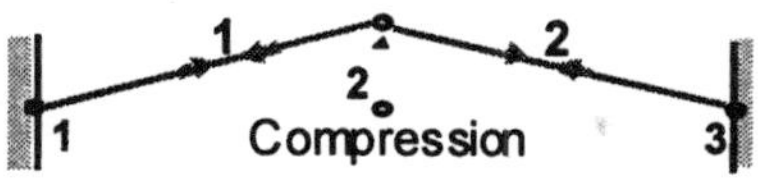

Figure 3.24 *Compression pre-stress*

In the first case, if the junction node is taken away from its initial location, the effects of the pre-stress of tension return it to this initial location: the infinitesimal mechanism is stabilised by the tension pre-stress.

In the opposite case, the set will find an equilibrium state for a different location, corresponding to a geometry satisfying:

E 3-5 $l_{ij} = d_{ij}$

The compression pre-stress does not return the system to its initial configuration.

The infinitesimal mechanism stabilisation by states of pre-stress (or self-stress) is an essential characteristic of tensegrity systems.

3-5. Conclusion

Fundamental concepts have been illustrated in this chapter – form-finding, pre-stress and self-stress states, infinitesimal and finite mechanisms, stabilisation of infinitesimal mechanisms. They constitute the mechanical background which are necessary prerequisites for the understanding of the mechanics of tensegrity systems. Tensegrity systems require a form-finding process to reach a null self-stress equilibrium state: in this geometrical state, a self-stress state can be introduced, finite mechanisms have disappeared, but occasionally some infinitesimal mechanisms remain. self-stress states may or may not stabilise these infinitesimal mechanisms; it is necessary to check this stabilisation. The whole set of concepts has been illustrated with a balloon analogy, the balloon being considered as one tensegrity system among others.

References

Ref 3-1. Motro R., Formes et Forces dans les Systèmes Constructifs, Cas des Systèmes Réticulés Spatiaux Autocontraints, Thèse d'Etat, Université Montpellier II, June 1983.

Ref 3-2. Vassart N., Laporte R., Motro R., "Determination of mechanisms's order for cinematically and statically indeterminate systems", *International Journal of Solids and Structures*, Vol. 37, 2000, pp. 3807–3839.

4

Typologies

4-1. Introduction

This chapter is devoted to the description of the typology of tensegrity systems. We are dealing here only with tensegrity systems that can be included in spatial structures and not with other tensegrity systems such as biological cells.

Typology criteria are numerous: topology, geometry, mechanical characteristics (in terms of self-stress states and infinitesimal mechanisms) etc. It would certainly be useful to develop studies enabling us to have some "tables", governed by one or more of these criteria. However, this is not the aim of this chapter; we are only dealing here with some of the better known and most explored tensegrity systems. If, until now, two kinds have been developed – namely cellular units (or elementary cells) and their assemblies – we know that current developments concern complex tensegrity systems without an identified constitutive cellular unit. That is why this chapter is more a historical typology study than a prospective one, since we think that design procedures will lead to new possibilities (some of which will be described in the Chapter 7). Nevertheless we open this chapter with some comments on the criteria of classification with numerous examples, followed by an attempt at codification for elementary cells. We will then give a description of elementary cells, assemblies of cells and also of some of the more "exotic" tensegrity systems. And we will conclude with our recent proposals concerning double-layer grids.

It is out of the scope of this study to give extensive charts of tensegrity, even if this might well appeal to some people, who might want to immediately become acquainted with topological and geometrical characteristics. As things stand at present it is not necessary to try to present an exhaustive list, since as is the case for other spatial systems, each designer is able to define new types. Nevertheless, it appears that in respect of tensegrity systems, it is also possible to find some known classification such as double layer grids, with or without curvature (single or double curvature, positive or negative), domes, masts etc.

Moreover, as we suggested in Chapter 2, it appears that a very simple classification is defined in relation to the concept of "discrete pneumatic structures". Consequently, the general class of tensegrity systems contains potentially tensioned components inside a convex tensional "discrete" net. Classical elementary cells, which are described below are a subclass with only one tensional net, and, without tensioned components, inside this net.

4-2. Typology criteria and codification

4-2.1. Topology, geometry and equilibrium

4-21.1. Topology

As previously introduced in the "extended" definition in Chapter 2, the topology of the two sets of components is clearly defined: a set of tensioned components is continuous, a set of compressed components is discontinuous. For the first set, some other properties arise – for instance in the case of the so-called "spherical cell" – for which constitutive elements are homeomorphic[1] to a sphere. The elementary topological characteristics are the number of nodes "n", the number of struts "S" and of cables "C". In the case of standard systems (verifying all the elements of the "historic" definition), C is equal to half of N, but this is not the case when some nodes receive just cables. It is worth stressing that graph theory is very useful to qualify and study the topology of tensegrity systems. The topology of the system is completely defined when the list of members and of their ends are given; these two lists constitute what can be referred to as the "relational structure" of the system [Ref 4-1].

As an example, for the "elementary equilibrium" described by David Georges Emmerich (Chapter 1), the relational structure is given in Table 4.1 and can be represented by the graph in Figure 4.1 (tension and compression refer to stresses in the corresponding elements of the simplex). It should be noticed that in this case another choice could have been made for graph links corresponding to struts: for instance node 3 may be linked with node 6 instead of node 5 (and similarly for other links). These two possible graphs are associated with two geometrical solutions according to the chosen rotation around an oriented vertical axis. These are known as "levogyre" (anticlockwise) and "dextrogyre" (clockwise).

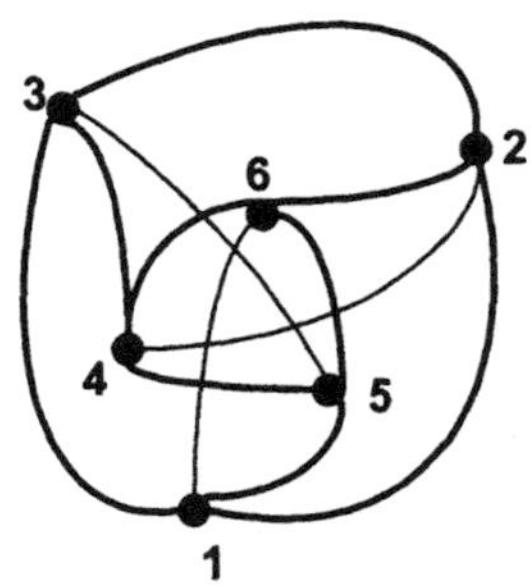

Figure 4.1 *Graph of elementary equilibrium*

[1] They can be applied onto a sphere without any cutting or overlapping of elements.

Table 4.1 *Nodes and elements of the elementary equilibrium*

Element	Type	End 1	End 2
1	Compression	1	6
2	Compression	2	4
3	Compression	3	5
4	Tension	1	2
5	Tension	2	3
6	Tension	1	3
7	Tension	3	4
8	Tension	1	5
9	Tension	2	6
10	Tension	4	5
11	Tension	5	6
12	Tension	4	6

4-21.2. Geometry

For a defined topology, geometry is characterised by coordinates x_i, y_i and z_i for the n nodes of the system. Length "s" of struts and length "c" of cables can be derived from the coordinates.

Generally, the designer makes a choice concerning values of lengths. Whatever the design process for a tensegrity system might be, two situations have to be considered:

- If all the elements, for each given set, have the same length in the system, s is unique and so is c. We chose to call "regular" the corresponding systems. Geometry is then dependent upon only one parameter, the ratio s/c[2].
- If there are several values s_s and c_c for the corresponding elements in a same system the qualification "regular" will be used[3].

If topology, "s" and "c" are defined, then the geometry is qualified by the whole set of coordinates, which is closely related to the self-stress equilibrium.

4-21.3. Equilibrium and form-finding

If topological characteristics mentioned in the definition of tensegrity systems are important specific attention should also be paid to another characteristic, namely the "self-stress" which stiffens tensegrity systems. We will define precisely the concept of self-stress in Chapter 5. But it is sufficient to know at this stage that a self-stress state is such that every element is under tension or compression (according to its nature), the whole system being in static equilibrium without any external actions (self-weight has to be neglected). Self-stress requires equilibrium of each node. In the case of a "standard" situation a node receives three cables and one strut. A

[2] Sometimes these systems are called mono parameterised systems.

[3] These are multi-parameterised systems.

necessary condition of equilibrium can be expressed in geometrical terms: the strut (compressed element) has to be inside the solid angle defined by the three cables (tensioned elements)[4] (see Figure 3.2, in the preceding chapter).

Initially, the geometry of regular polyhedra was used to generate tensegrity systems. But misunderstandings have been made regarding equilibrium conditions. If we take a look at the truncated tetrahedron, for instance, it has been claimed that its geometry, when it is defined by its node coordinates, was consistent with the possible equilibrium of a tensegrity system. This system was built by inserting struts inside the cable net that is generated by the edges of this semi-regular polyhedron [Ref 4-2]. Simple considerations about equilibrium of one node are sufficient to establish that the truncated tetrahedron cannot be in equilibrium in its initial geometry (see those paragraphs in Chapter 5 dealing with the form-finding process).

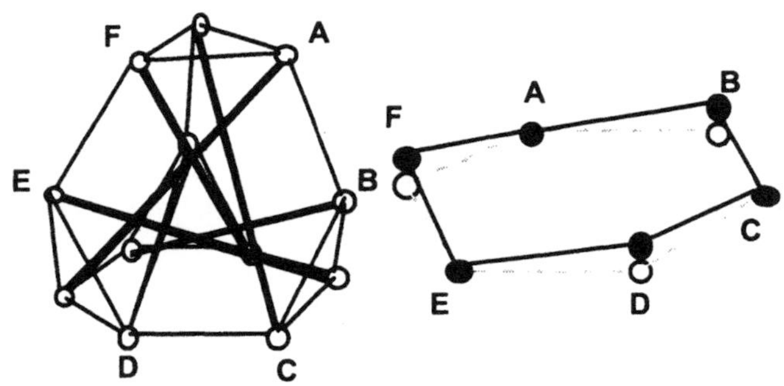

Figure 4.2 *Truncated tetrahedron*

4-2.2. Codification

It is always a difficult task to submit a codification. Several known tensegrity systems received a name, and sometimes more than one name (as was the case for the elementary equilibrium previously described). No doubt these names will remain, and quite rightly too. Nevertheless, we hereby submit a codification, which could be used alongside the accepted terminology, and for some cases it will be a useful tool to identify a tensegrity system. This codification is based on the subsequent ordered parameters:

- number of nodes "n";
- number of compressed components "S";
- number of cables "C";
- regular system "R" or irregular system "I" (according to the element lengths);
- spherical systems "SS" (according to the fact they fit a set of tensile components which is homeomorphic to a sphere, and corresponds to a single tensile discrete envelope).

[4] For simplification of expression, "strut" will be used for compressed elements and "cables" for tensioned elements if there is no risk of misunderstanding.

As an example, the code for the elementary equilibrium would be "**n6-S3-C9-R-SS**". Two tensegrity systems may have the same number of nodes and different numbers of cables (according to their topology). When there is one (and only one) strut at each node, the value of n is the double that of S. This codification could be reduced for regular, standard systems to just S-C values (S3C9 for the elementary equilibrium), or enlarged in order to provide more information (like the "levogyre" or "dextrogyre" qualifications). We will develop this codification further in later sections of this chapter.

4-3. Elementary or "spherical" cells

The definition of "elementary cell" is consistent with the concept of "single tensile discrete membrane" as previously mentioned; 'spherical cell' could be understood as an "indivisible" unit as is the case for an atom. It was mostly Emmerich ([Ref 4-3] [Ref 4-4]) and Pugh [Ref 4-5], (concerning the description of elementary cells) who have carried out a large amount of work. Their explanations are both topological (especially Pugh, who established families based on topological considerations) and geometrical. It is a fact that they made extensive use of polyhedral geometry and did not give many precise mechanical derivations that could explain, for instance, the misunderstanding concerning the previously described truncated tetrahedron.

4-3.1. "Spherical cells"

4-31.1. Characteristics

A first class of tensegrity systems may be defined according to a topological property associated with cables: the cable set is always homeomorphic to a sphere. When all cables can be mapped on a sphere without intersections between them, apart from the nodes of the system, then the cable set is "homeomorphic" to a sphere (Figure 4.3)[5]. All the struts are inside this cable net and the corresponding tensegrity systems are called "spherical cells".

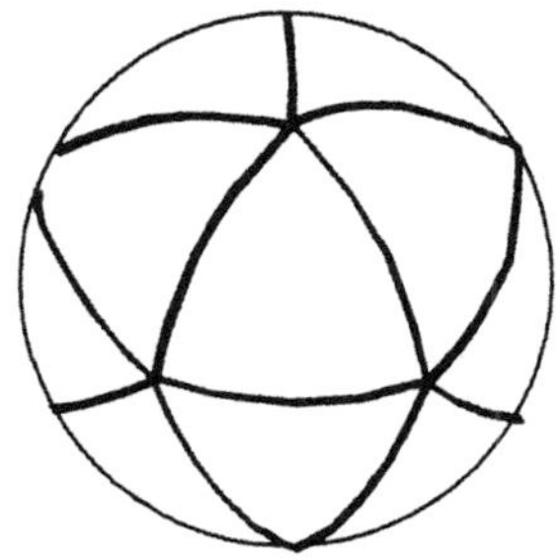

Figure 4.3 *Cable set mapped on a sphere*

[5] In this case, it can be demonstrated that the cable graph is plane. [Ref 4-1]

Figure 4.4 *Spherical tensegrity (Sanchez Cuenca)*

Most often described tensegrity systems are spherical cells. We shall follow, for the subsequent presentation, the topological classification given by Pugh: *rhombic*, *circuit* and *Z- configurations* are the three main classes.

4-31.2. Rhombic configuration and prismatic cells

4-312.1 The "Simplex" (elementary equilibrium)

Let us consider a straight prism with a triangular base; its edges are cables. One strut is inserted diagonally to each square face (Figure 4.5). A simple way to attain the corresponding tensegrity system (Figure 4.6) will be to operate a relative rotation between the upper triangle and the lower triangle. This tensegrity system is the smallest spatial one that can be built according to the so-called "patent" definition.

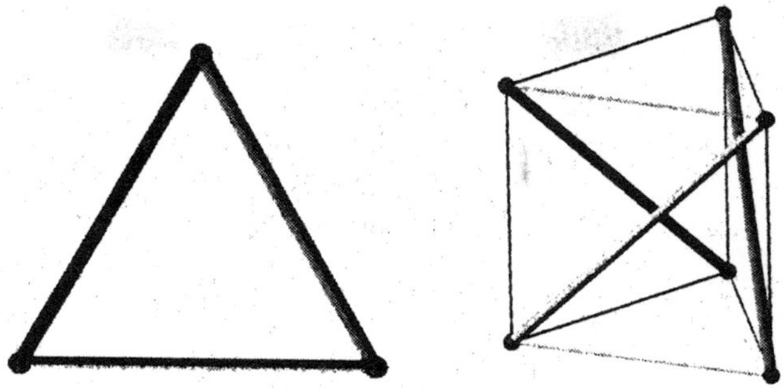

Figure 4.5 *Triangular prism*

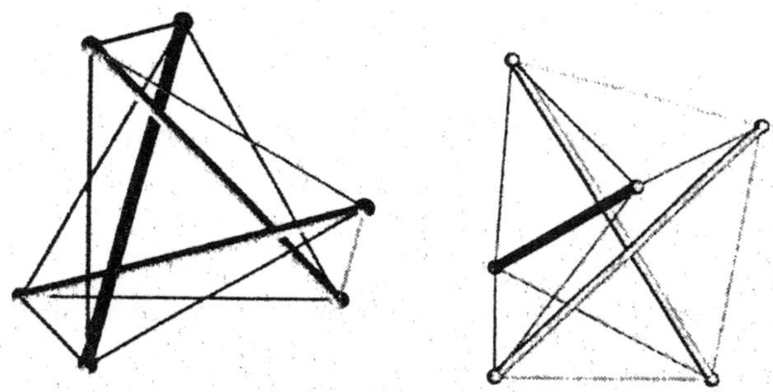

Figure 4.6 *Simplex*

The generation mode is the basis of the chosen denomination "prismatic system". As previously described, this system also received other names: *simplex, elementary equilibrium, twist unit, regular triplex etc.* Its codification could be n6-S3-C9-R-S or, more simply put in this case, S3-C9.

The Simplex has a rotation symmetry axis of order 3; then two enantiomorphic [Ref 4-3] varieties exist, one the "levogyre" simplex, the other the "dextrogyre" simplex. These terms are defined in connection with the relative rotation wise of the two basis triangles, when referred to the oriented rotation axis mentioned in Figure 4.7. The relative rotation angle is equal to 30°. Its value can be calculated by a form-finding process[6] and is an important characteristic of the Simplex; A. Pugh called it "twist angle" and G. Minke uses the denomination "twist unit" for this tensegrity system [Ref 4-7].

[6] A demonstration is given in Chapter 5.

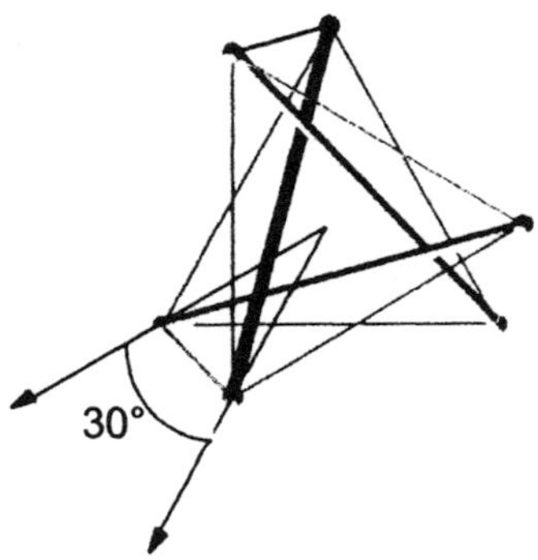

Figure 4.7 *Twist angle*

4-312.2 Prismatic tensegrity systems and their derivatives with "rhombic" configuration

Prismatic tensegrity systems

The principle described for the Simplex can be extended to other prisms, with square, pentagonal, n-gonal... bases. Under some conditions a class of so-called "prismatic" Tensegrity systems can be defined. Equality of cable lengths is by no means guaranteed for all of these systems [Ref 4-5].

In this particular class, the struts and cables (which are called "bracing cables"), which do not constitute the base, can be considered as generators of two revolution hyperboloïds H_s, H_c. They intersect each other along two circles, which contain the corresponding polygonal base. Each of them possesses two generator families that correspond to the two enantiomorphic varieties, which have been already mentioned in respect of the Simplex (Figure 4.8).

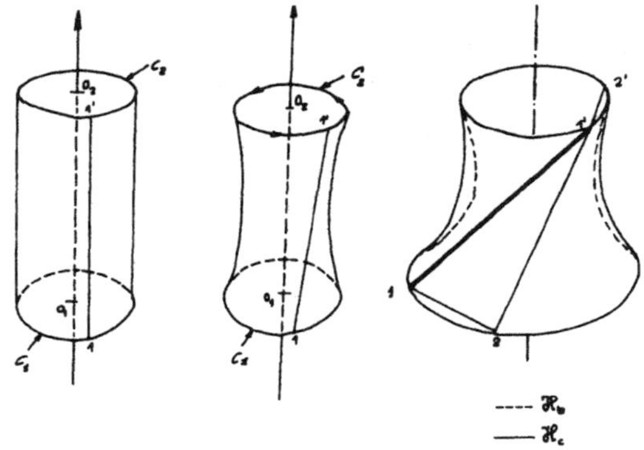

Figure 4.8 *Generating hyperboloïds*

Rhombic configuration

In the classification recommended by A. Pugh (who worked in collaboration with R.B. Fuller) the total prismatic series belongs to a larger class of so-called "rhombic systems" (or "diamond systems"), which is characterised by a rhombic configuration defined as follows by Pugh:

"Each strut of a "rhombus system" constitutes the longest diagonal of a rhombus of cables, folded according to this axis" (Figure 4.9).

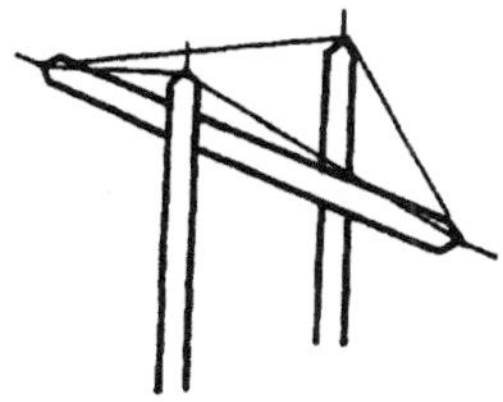

Figure 4.9 *Rhombic configuration*

The prismatic class satisfies this definition and each prismatic tensegrity system comprises a single layer of struts (the word "layer" is in fact linked to modes which were used by Pugh to build his models). Other systems can be designed by multiplying the number of layers. It is then necessary to superpose several prismatic tensegrity systems and to satisfy certain conditions (especially those struts included in a new layer that have one extremity localised at the middle point of the base cables of the previous layer: this is a strut on cable junction) (Figure 4.10) ([Ref 4-8] and [Ref 4-9]).

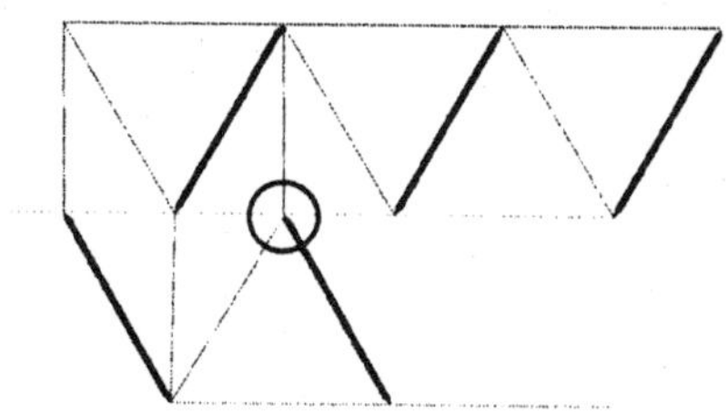

Figure 4.10 *"Strut on cable" junction*

Table 4.2 summarises the classification proposed by A. Pugh for tensegrity systems with "rhombic configuration". Pugh's view is that the deformability of the resulting model increases with the number of layers. But this characteristic requires a more accurate mechanical study.

Table 4.2 *Multi layer tensegrity systems with "rhombic configuration"*

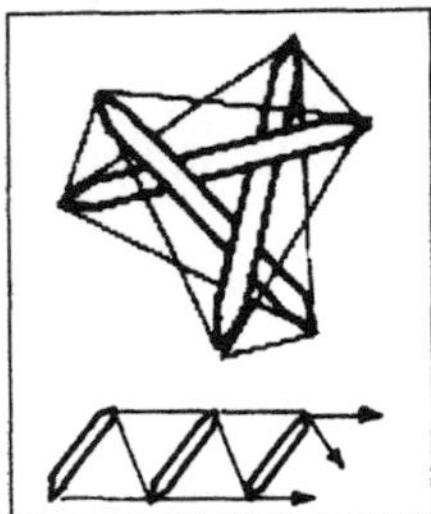	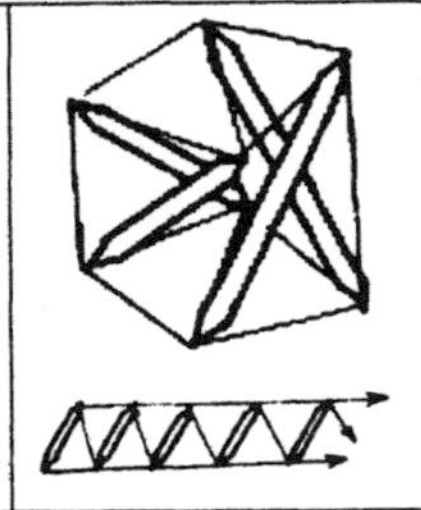	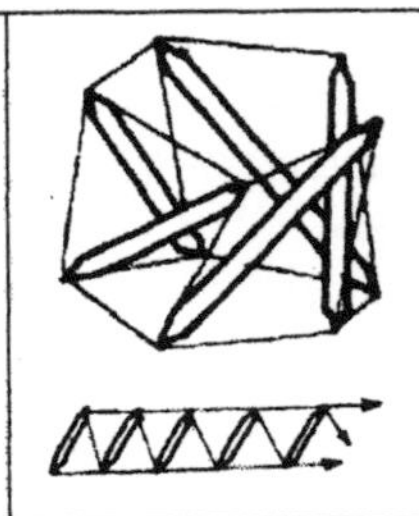	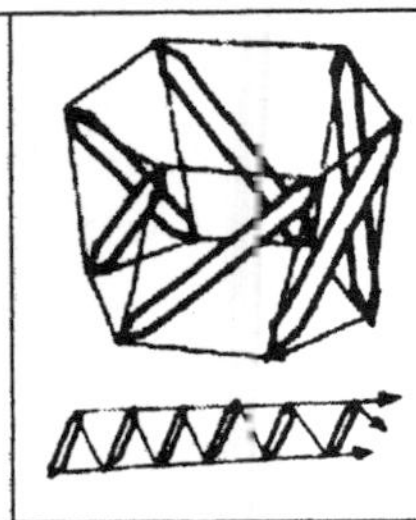
n6S3C9	n8S4C12	n10S5C15	n12S6C18
Single layer tensegrity systems ("prismatic")			
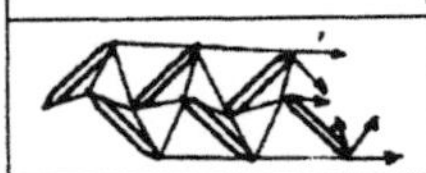	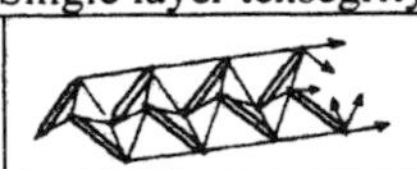	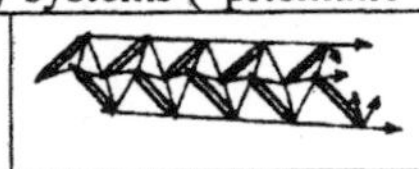	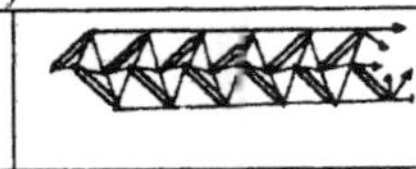
n12S6C24	n16S8C32	n20S10C40	n24S12C48
Double layer tensegrity systems			
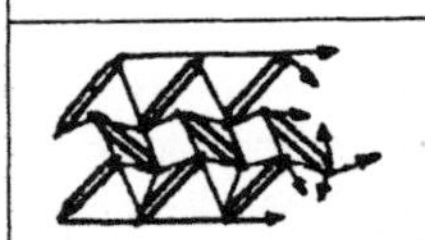	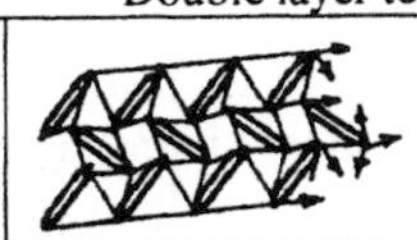	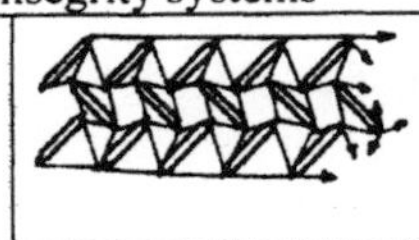	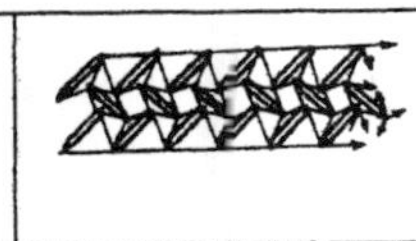
n18S9C36	n24S12C48	N30S15C60	n36S18C72
Double layer tensegrity systems			

Tensegrity systems are classified in this table according to the number of "layers", the simplest one being the simplex, which has already been described in some detail in this book. It is also worth emphasising another "popular" tensegrity system, which comprises two layers. This system can be achieved by the association of two simplex ("node on cable" junction) by means of six extra cables. Six struts and twenty-four cables constitute this system (codification n12-S6-C24-R-SS, or more simply in this case S6-C24). It will be noted that the struts are parallel to one another, two by two. If "s" is the strut length, it can be demonstrated that the equilibrium geometry is attained when the minimum distance between two parallel struts is equal to s/2. The corresponding value of the ratio s/c is then r = 1,67. Its usual name, for some, is the "icosahedric tensegrity system", but if there are indeed twenty triangular faces on its outside, all these faces are not identical. Nor are they all completely closed by a set of cables. Nevertheless, in this case the rhombic configuration is clear. We ourselves used a similar denomination combining Emmerich's rule with the icosahedric property, since we referred to it as "*autotendant icosaédrique*" in a research paper related to its equilibrium[7]. But some years after we chose another term "expanded octahedron" (Figure 4.11): in fact the two-layer system studied in these lines can be considered as the result of a splitting operation of the octahedron diagonals. H. Kenner [Ref 4-10] described some interesting results concerning this system, that are related to its geometry and a so-called "elasticity multiplying factor": the mechanical behaviour of this system is

[7] Corresponding demonstration is given in Chapter 5.

such that when two parallel struts are discarded one from one another, the two other pair of struts are also discarded.

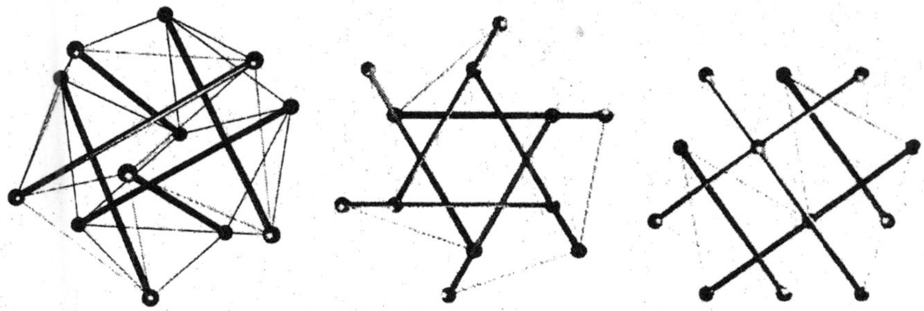

Figure 4.11 *Expanded octahedron*

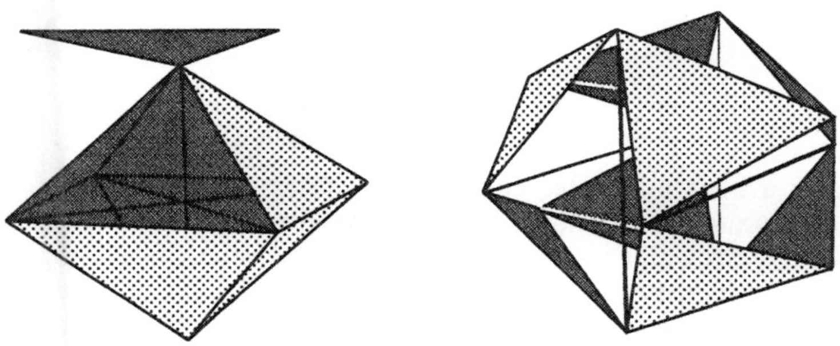

Figure 4.12 *Splitting the diagonals of an octahedron*

4-31.3. "Circuit" cells

<u>4-313.1 Definition and characteristics</u>

This second class is characterised by the existence of strut circuits[8]. Circuit cells do not strictly satisfy the "patent" definition given for tensegrity systems: in this case "compressed" elements are polygons of struts, which have to be considered as compressed components. Their construction can, for example, derive from the prismatic class by completely closing the rhombus of cables: the ends of two struts are joined (according to the shortest diagonal of the rhombus). Consequently, the number of cables is then divided by two, since they had soon a common node on the red strut (Figure 4.13).

---

[8] In graph theory, a circuit is a path whose two ends are merged in one apex. A path is a number (non equal to zero) of arcs so that the final end of each arc coincides with the initial end of the following arc. All the apices of the path are not necessarily separate.

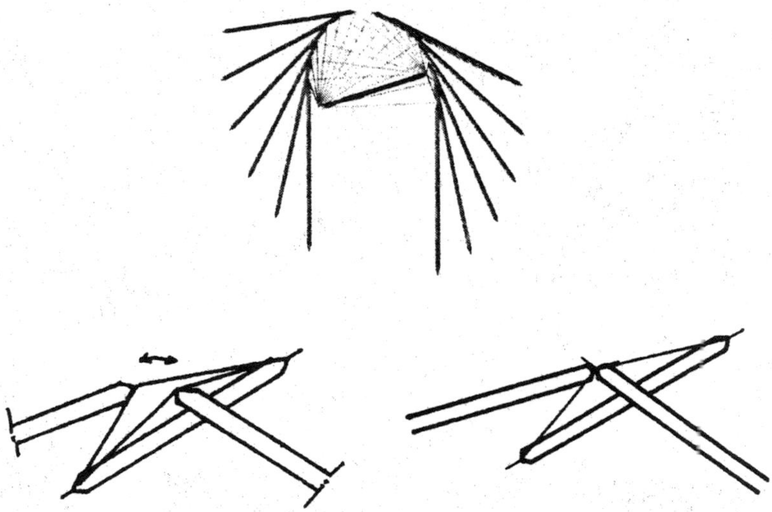

Figure 4.13 *"Circuit cells" principle*

Thus taking as a basis a rhombic configuration system with three layers, each comprising four struts, the described transformation gives birth to a system whose cable set geometry fits with a cuboctahedron; the struts constitute four triangles that are the circuits of this system (Figure 4.14). The codification[9] of this system expresses its properties: two struts at each node implying an equal number of struts and nodes. This system is regular but not standard, since the compressed elements are not single straight struts, but constitute compressed components: at every point matter is under a compression stress.

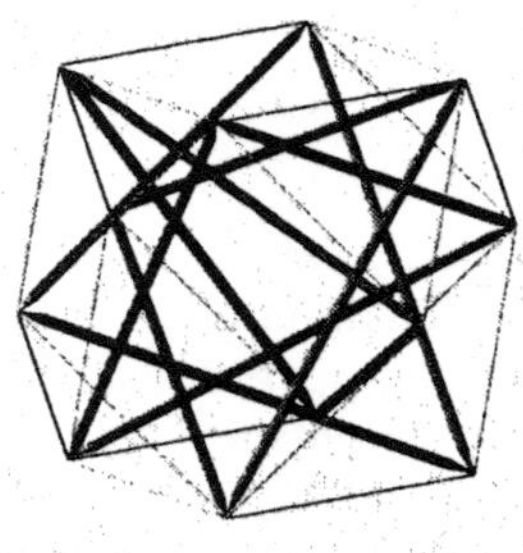

Figure 4.14 *Cuboctahedric circuit tensegrity systems*

The intertwining of circuits was inscribed in the name "basketry tensegrity", which was originally used by R.B. Fuller. Indeed, corresponding geometries have nothing to match the subtlety of basketry work.

[9] Codification n12-S12-C24-R

Besides the existence of circuits, an essential characteristic of these systems is the number of cables (4) necessary to ensure the stability of a node. On the basis of this characteristic, other tensegrity systems with circuits can be derived from known polyhedral geometries. The number of circuits can even be reduced to a single figure as I had the opportunity of checking on a real model recently.

According to Pugh's experience, the models of these systems are less flexible than rhombic systems possessing the same number of layers; this can easily be understood since they are derived from them by a contraction.

4-313.2 Regular and semi-regular polyhedra related to circuit tensegrity systems

Among the regular and semi-regular polyhedra [Ref 4-11], five have apices with four edges and can serve as a geometrical basis to constitute circuit tensegrity systems.

The characteristics of the corresponding tensegrity systems are given in Tables 4.3 to 4.5.

The octahedron has to be considered as a borderline case, for which circuits are reduced to a single strut lying on each diagonal. The three diagonals intersect each other at the centre of the circumscribed sphere. One could then consider this system as belonging to a different class whose struts would be in the shape of star components, like the elements of Maraldi chains. This figure is nevertheless interesting since when considering the inverse transformation that which has allowed the transformation of rhombic systems into circuit systems, the octahedron gives birth to the system previously referred to as expanded octahedron (see Figure 4.11). The chosen name can be understood on this basis.

Table 4.3 *Circuit tensegrity systems based on octahedron and cuboctahedron*

a- octahedron Gives access to the so-called expanded octahedron.	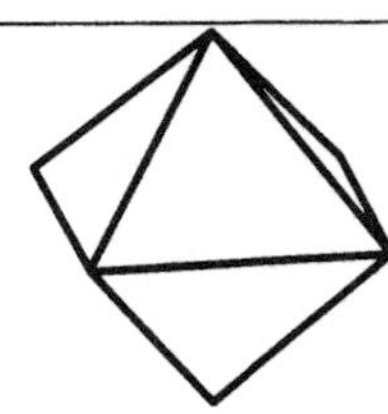
b- cuboctahedron A tensegrity system with four quadrangular compressed components. The twenty-four cables lie on the edges of the cuboctahedron.	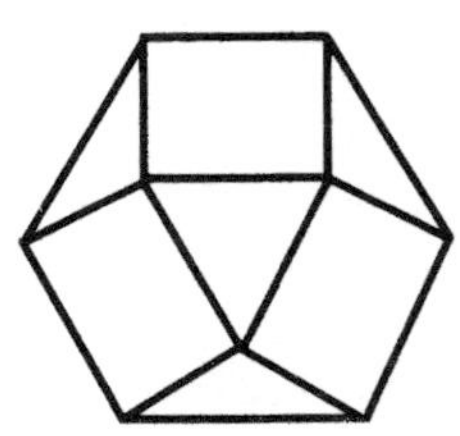

Geometries b) and c) correspond to equilibrium geometries of circuit systems. The polyhedron, which is linked to the cube and to the octahedron families, has been at the centre of R.B. Fuller's preoccupations elsewhere – under the name "Dymaxion" [Ref 4-12].

Table 4.4 *Circuit tensegrity systems based on icosidodecahedron*

c- icosidodecahedron

Six compressed components, with pentagonal shapes, are inserted in a set of sixty cables, each one being an edge of the regular icosidodecahedron.

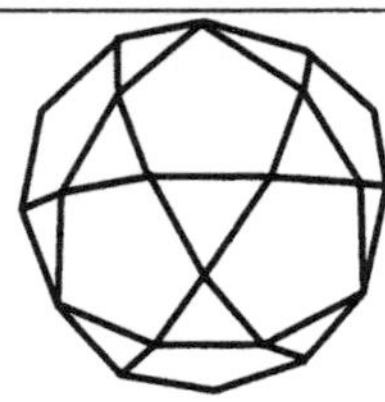

Table 4.5 *Circuit tensegrity systems based semi-regular polyhedra*

d- small rhombicuboctahedron

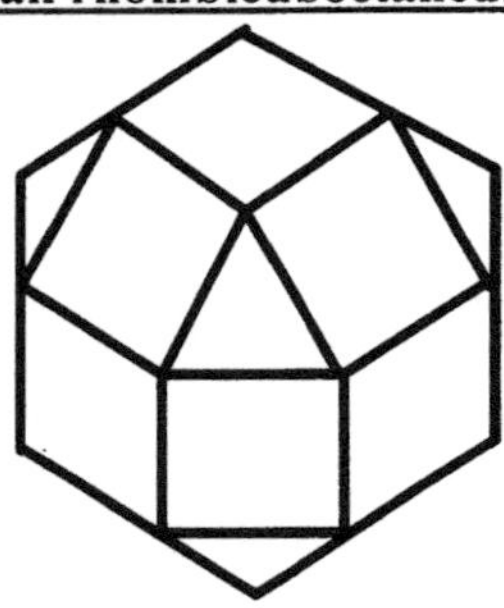

Six compressed components with square shapes are inserted in a tensile continuum resulting in equilibrium shape, which fits with the snub cube.

Snub Cube

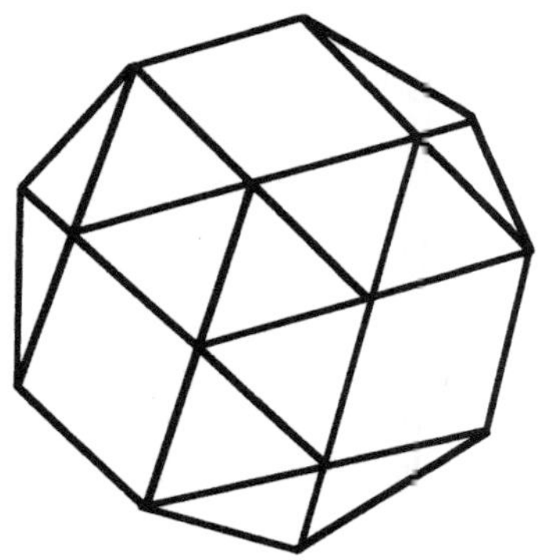

e- small rhombicosidodecahedron

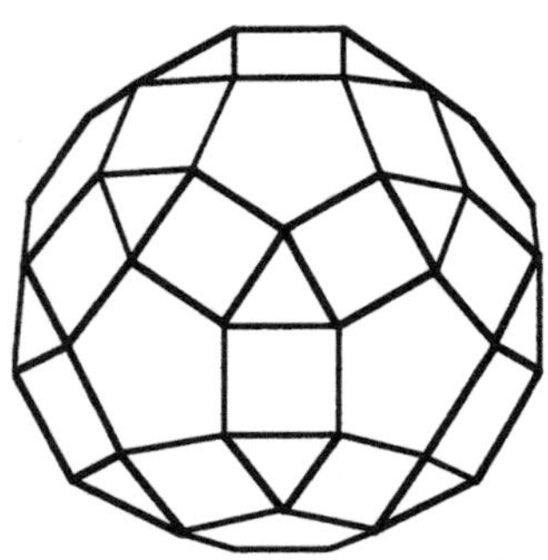

The small rhombicosidodecahedron produces a tensegrity system with twelve pentagonal compressed circuits, whose edges constitute a snub icosahedron.

Snub Icosahedron

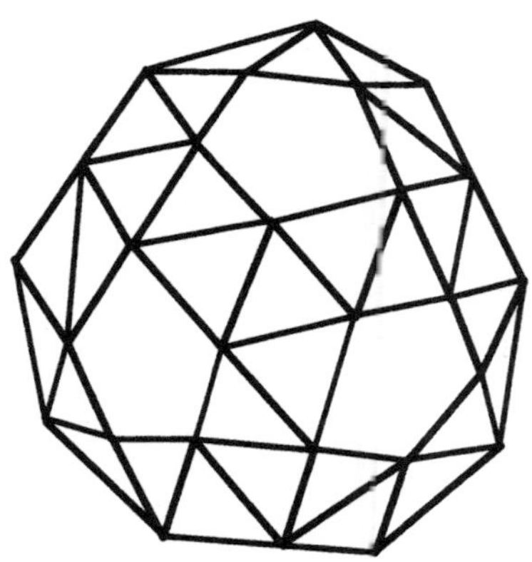

With polyhedra d) and e) equilibrium can not be reached in the initial geometry since the circuits are such that the incidences of struts ending at the squares' apices are not identical and induce the distortion of these faces.

Two solutions are then possible:

- In order to maintain the initial geometry, complementary cables can be added; they stabilise the squares.
- The equilibrium position of these systems can be attained in a geometry which is different from the initial one. In this second hypothesis the small rhombicuboctahedron leads to the geometry of the snub cube, which would lack some edges (corresponding drawing in Table 4.5 comprises all the edges (5) of the snub cube, four of them are only necessary for equilibrium). The small rhombicosidodecahedron finds its equilibrium geometry according to snub dodecahedron geometry. It is necessary to note that two enantiomorphic figures exist for the snub cube and the snub dodecahedron depending upon the choice made for introducing circuits in the initial shape.

<u>4-313.3 "Geodesic" tensegrity systems with circuits</u>

Under the decisive impetus of R.B. Fuller, the development of geodesic domes has witnessed a significant improvement during the second half of the twentieth century. The mode for their geometric generation is described in [Ref 4-13]; it is sufficient here to bear in mind that this generation relies on the division of triangular faces (sometimes square faces) of the polyhedra which are chosen to generate the dome. The breakdown in triangles of a basic one can be realised according to two modes. It is a "Class 1 breakdown" which is normally used [Ref 4-10]. Fuller qualified it as an "alternate breakdown".

Besides the class of chosen breakdown, it is the number of generated segments on an edge of the initial polyhedron which characterises the geodesic breakdown. This number is called "breakdown frequency" (Figure 4.15).

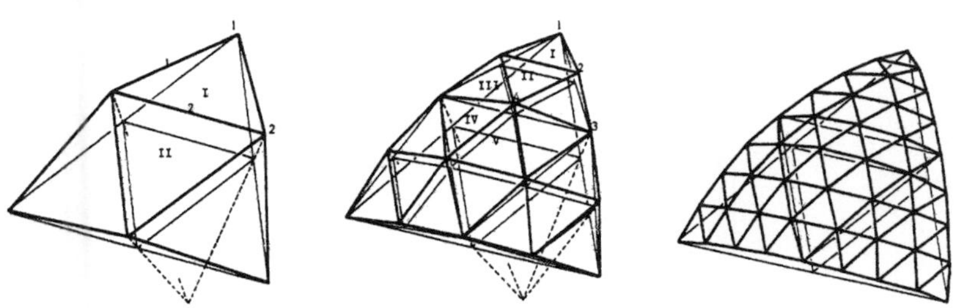

Figure 4.15 *Breakdown frequency*

The examination of geodesic geometry shows that for Class 1 with even breakdown frequencies, the resulting systems can be compared with tensegrity systems with circuits. Indeed (Figure 4.16), the system in heavy lines possesses the characteristics of circuit systems; one finds four connections for each node. The configuration principle of these tensegrity systems is illustrated in Figure 4.17. The intertwining of struts that belong to circuits is quite clear there.

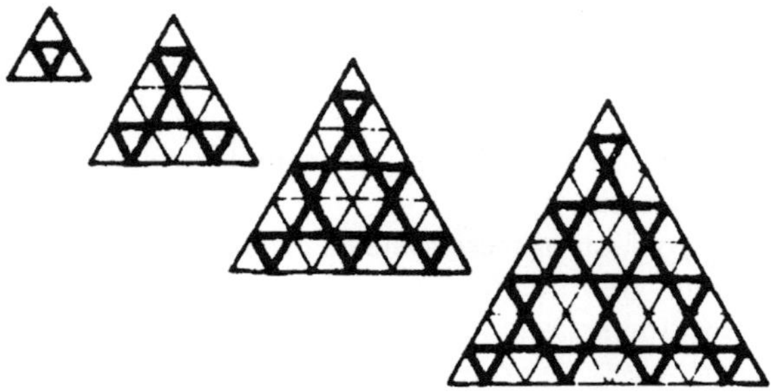

Figure 4.16 *Distribution of struts and cables on a geodesic pattern*

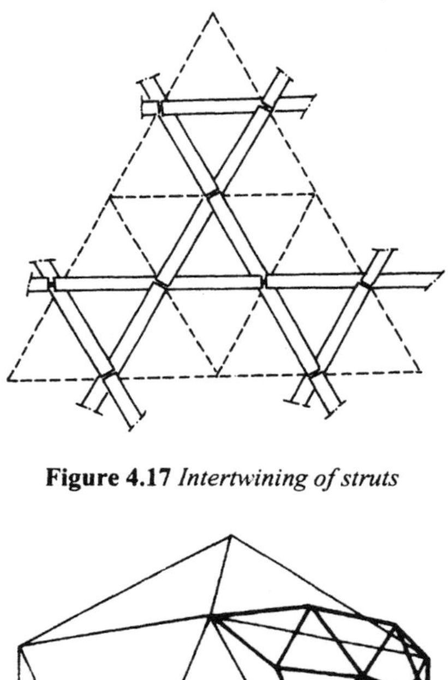

Figure 4.17 *Intertwining of struts*

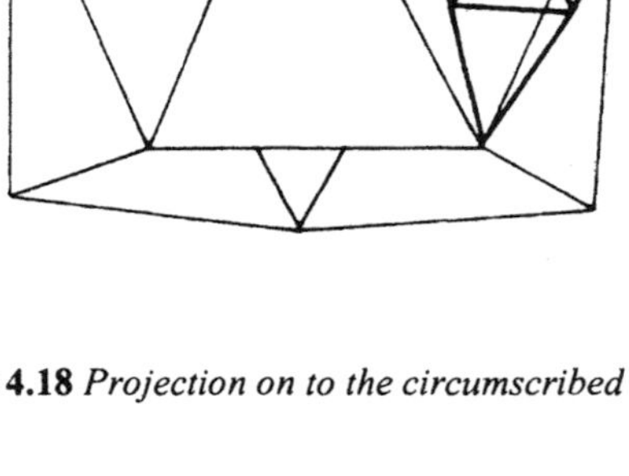

Figure 4.18 *Projection on to the circumscribed sphere*

It is necessary, as in the case of classic geodesic domes, to project the geometry of the generic polyhedron on to its circumscribed sphere. In fact, only nodes have to be projected (Figure 4.18).

4-31.4. Type Z tensegrity systems

<u>4-314.1 Definition and example</u>

The definition given by A. Pugh for this new class is also of a qualitative order:
"A type Z 6 tensegrity system (or 'Zig Zag' tensegrity system) is such that between the two extremities of each strut there exists a totality of 3 non aligned cables". These three cables then do indeed form a "Z". These systems can be constructed when taking as a basis "rhombic" systems according to the principle of Figure 4.19. This diagram shows the existence of two possibilities. When the transformation is operated on the totality of a tensegrity cell, the coherence has to be preserved when choosing the suppression of the two cables.

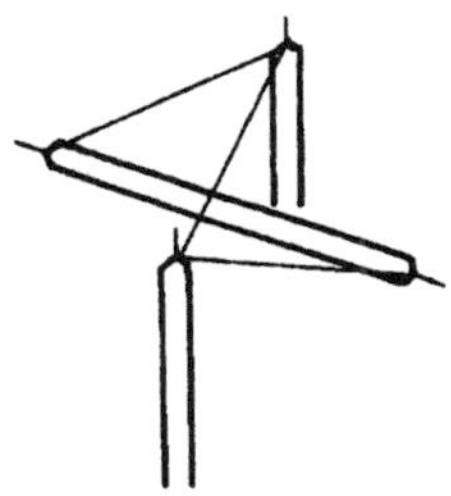

Figure 4.19 *Zig Zag connection principle*

Thus the transformation of the expanded octahedron leads to a geometry which is close to the truncated tetrahedron geometry (Figure 4.20).

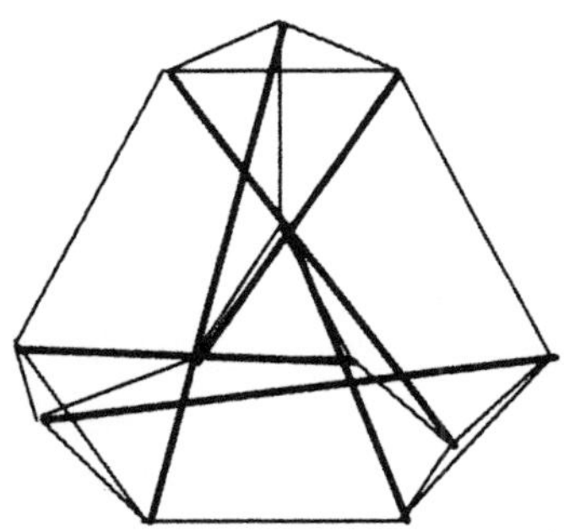

Figure 4.20 *Truncated tetrahedron*

In this type of tensegrity system one notes there are three connections by nodes, and this is a point worth bearing in mind in respect of the following discussion.

4-314.2 Regular and semi-regular polyhedra in connection with type "Z" tensegrity systems

One of the characteristics of "Z tensegrity systems" is that they comprise 3 connections by node. Three platonic polyhedra and seven archimedian polyhedra satisfy this condition.

Although Pugh did not note it specifically, this does not give any certainty of attaining a balanced geometry. It is a point which we will discuss further in the chapter dealing with models. The cube and the octahedron examples are the converse demonstration: one cannot obtain with the application of the "Z" transformation balanced figures, which strictly maintain the initial geometry.

4-314.3 Geodesic "Z" tensegrity systems

Following the same process as in case of geodesic "circuit tensegrity systems", the use of geodesic breakdowns allows us to extract meshes that are necessary for the constitution of "Z" tensegrity systems. For this, it is necessary for the breakdown frequency to be a multiple of 3. Figure 4.21 indicates how the existence of 3 connections for each node of systems in heavy lines is likely to generate the required systems.

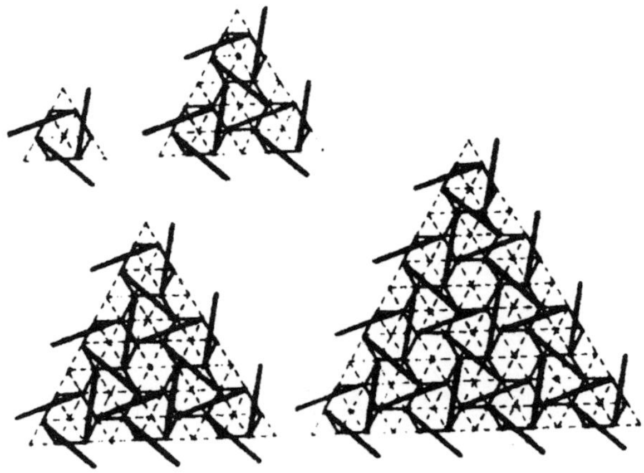

Figure 4.21 *Geodesic "Z like" patterns*

Hexagons of the mesh can be classified in two categories a) and b) (Figure 4.22) according to the incidence of struts ending at their apices. Incidences of type b) are such that the angular hexagon regularity cannot be preserved – which explains the distortions that occur while building models.

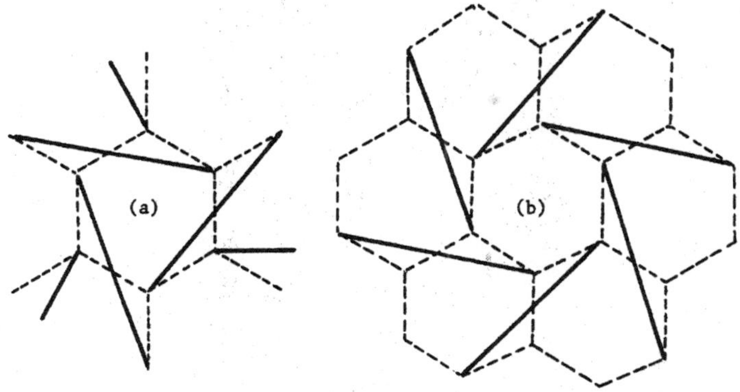

Figure 4.22 *Two different hexagonal meshes*

4-3.2. Stars cells

Some derivations may be made taking spherical cells as a basis. As examples we can quote the so-called "stars cells" designed by V. Raducanu [Ref 4-14].They constitute an extension of four strut prismatic tensegrity cells. Their diagonals, respecting equilibrium considerations replace upper and/or lower cable squares. Three possibilities are defined (Figure 4.23).

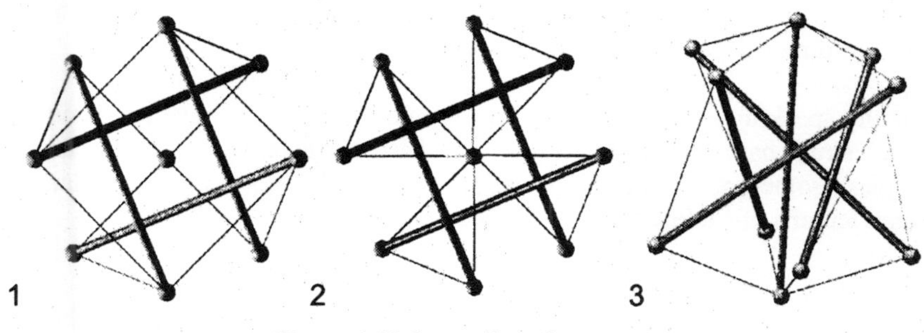

Figure 4.23 *Star cell configurations*

For case 3, it can be noted that some nodes receive only two tensile elements and one strut, and these three elements are coplanar. The spatiality of cells does not require the existence of three cables at each node as some have claimed in the "patent" definition. Furthermore, tensegrity systems can be established with some nodes without struts (see Figure 4.24).

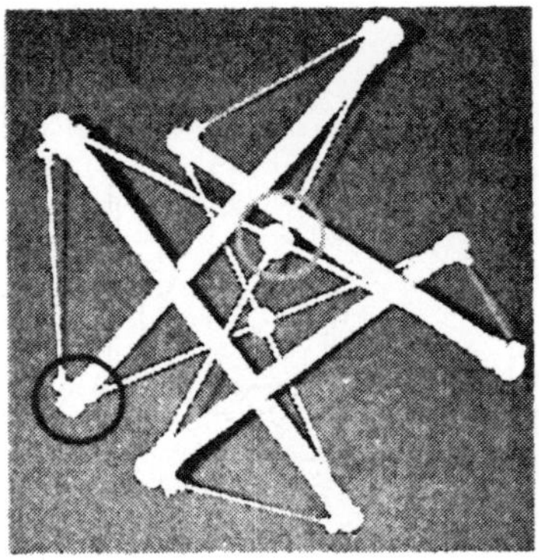

Figure 4.24 *Star cell with specific nodes (Raducanu, V.)*

4-4. Assemblies of cells

4-4.1. Introduction

It was entirely natural to try to assemble elementary cells to constitute more complex tensegrity systems. The following descriptions are, once again, based only on geometrical considerations, even if, sometimes, equilibrium is maintained. It is not worth providing an exhaustive list of proposals that have been made in this field. We only quote here some significant examples, which can be classified in three kinds: uni-dimensional, bi-dimensional and three-dimensional systems.

Generally speaking elementary cells are "regular": All cables have the same length "c", all struts have the same length "s". This is not always true for assembly examples: sometimes, assemblies can only be achieved with "irregular" cells (several lengths of cables, several lengths of struts).

4-4.2. Uni-dimensional systems

These systems are characterised by a predominant axis, which dictates the whole geometry.

This class is illustrated by physical models, and/or sculptures which have been created mainly by Snelson, Fuller and Emmerich.

4-42.1. Tensegrity masts

According to the chronology of events, Snelson and Fuller proposed a tensegrity mast at precisely the same time – a matter of dispute, however, between the two protagonists themselves. In his book "The Dymaxion World of Buckminster Fuller", [Ref 4-12] the author actually presents a picture of a Tensegrity Mast made by Kenneth Snelson, 1949 (Figure 4.25). On the same page another mast is presented. This mast (the legend is Tensegrity Mast, North Carolina State College, 1950, Figure 4.26) seems to have been designed by Fuller – as was a third one (University of Oregon Tensegrity Mast, 1953, Figure 4.28). The second one (or a similar) was exhibited for the first time at the Museum of Modern Art, in New York, and recently in Paris for the exhibition the "Art of Engineering" (Figure 4.27).

Figure 4.25 *Fuller and Snelson mast*

Figure 4.26 *Tensegrity mast, North Carolina State College, 1950*

Figure 4.27 *Tensegrity mast: detail*

Figure 4.28 *University of Oregon tensegrity mast, 1953*

It is interesting to establish a relationship between these masts and the early sculptures of Kenneth Snelson. The compressed components are made with two opposite "V" shapes in two perpendicular planes, which can be understood as an evolution of the initial double "X" submitted by Snelson, with an appropriate set of cables. Moreover, these tensegrity systems have compressed components, which are not simple straight struts; this remark is made in relation with the submitted extended definition previously given. The last comment concerns the detail in Figure 4.27: the two "V" are separated, each one contains two struts assembled with a spherical node, and the two nodes are joined with a small straight tensioned component. The spirit of tensegrity is included in these early examples.

4-42.2. Needle Tower

The Needle Tower (Figure 4.29) was designed by K. Snelson among several other tensegrity sculptures. It came to my own attention when exhibited in the gardens of Hisshorn Museum. It is more elaborate than the initial tensegrity masts and is certainly one illustration of the aesthetic qualities of tensegrity systems, in keeping with other Snelson sculptures.

Figure 4.29 *Needle Tower geometry*

If there was no chronological precedent, one could have told that the mast designed by K. Snelson improves others by complementary cable addition. Its system is in fact a "rhombic" tensegrity system with several layers of three elements each. It can also be considered as a superposition of expanded octahedra.

4-42.3. Emmerich's proposals

Two other masts are the result of work undertaken under the direction of D.G. Emmerich.

The "Chinese mast" of P. Boulet is built by the direct assembly of nodes of successive "simplex" [Ref 4-15]. The so-called "Mât Autotendant" "*Mât Autotendant*" by A. Chassagnoux necessitates the junction of lower nodes on the middles of cables of the previous one in the mast (Figure 4.30). This mast was the object of a physical test in Strasbourg [Ref 4-16]. Its constitution itself gives a very flexible system.

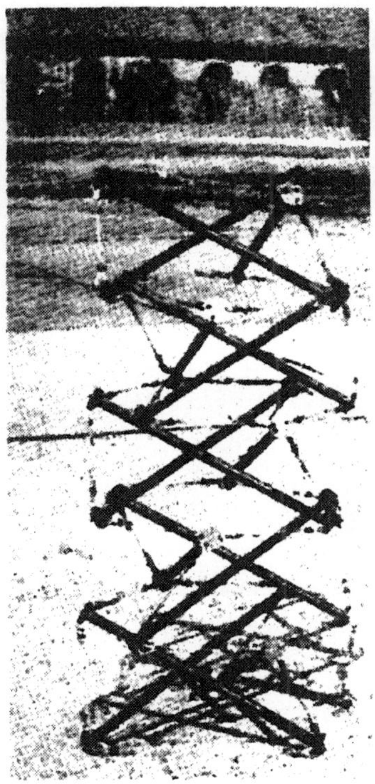

Figure 4.30 "*Mât Autotendant*"

Still in the field of uni-dimensional assemblies, a "*torus*" (Figure 4.31) designed by two other students of D.G. Emmerich preserves the principle of simplex junction end-to-middle of cable, but the authors closed the system and thus created a system whose connectivity is of a superior order. This system constituted a first step of topological possibilities of assemblies. Not all the cables have an identical length.

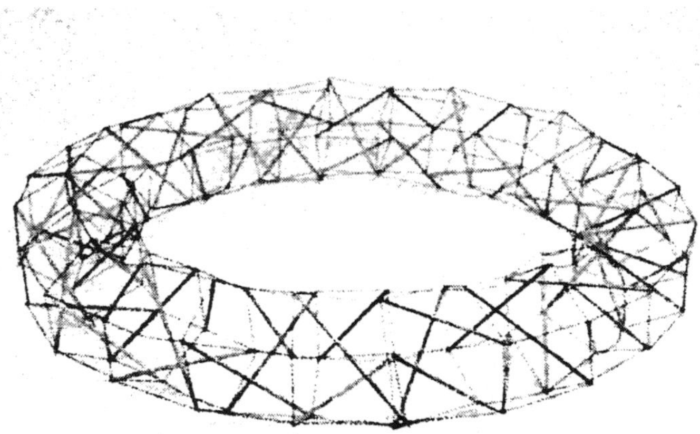

Figure 4.31 *Tensegrity "torus"*

We also experimented, in 1999[10], with some other uni-dimensional shapes, such as the arch of Figure 4.32 using two straight cables to balance the thrusts.

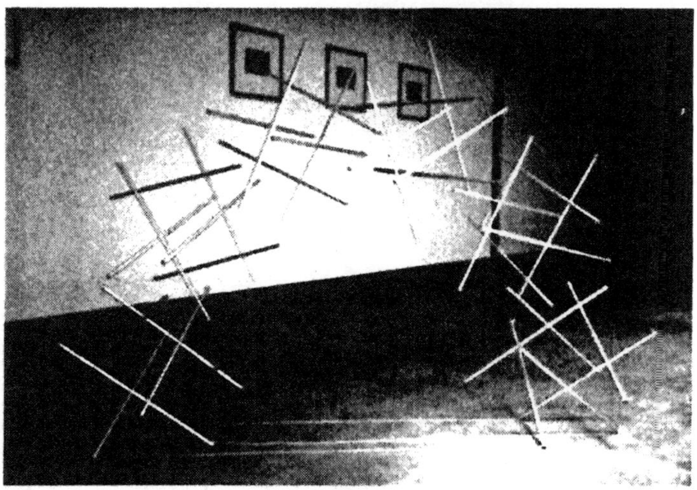

Figure 4.32 *Arch with six expanded octahedra*

4-4.3. Bi-dimensional assemblies

Examples of bi-dimensional assemblies are described in this section. It is still in the scope of tensegrity systems resulting from the assembly of elementary self-equilibrated cells. Some innovative designs have been developed in recent years and their principle is described in Chapter 7. The design of nodes and components are not described in this chapter, which deals only with typologies. Several junction modes can be used: node on node, node on cable, cable on cable (partial), cable on cable (total). Hanaor [Ref 4-17] described the last three solutions. In type "1"

[10] Workshop on tensegrity, *Ecole Spéciale d'Architecture,* Paris, May 1999.

(Figure 4.33) the junction is operated with a single bracing cable: each of its two ends lies in different horizontal planes, and Hanaor used this mode in a plane configuration leading to a double layer grid (Figure 4.34).

Figure 4.33 *Junction modes*

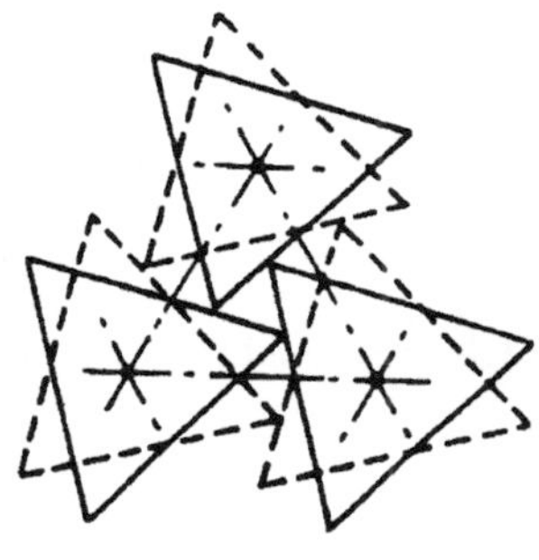

Figure 4.34 *Three cells junction*

4-43.1. Plane double layer tensegrity grids

<u>4-431.1 Strut on cable solutions</u>

The simplest idea is to assemble elementary cells so as to constitute a double layer grid, as can be done in a tiling operation. Elementary cells are self-equilibrated and so is their assembly. In some cases it is necessary to add additional cables in order to make the connections between cells. Elementary cells with three and four struts are commonly used. The relative rotation of upper and bottom polygon of cables only conditions the self-equilibrium of each cell.

Hanaor and Kono submitted their own proposals. Others can be found in Emmerich's works. But Emmerich did not give any mechanical support for his grids. Hanaor built prototypes, which are described in the available literature. [Ref 4-17]. Kono published a study on the dynamic behaviour of its grid [Ref 4-17] (see Figure 4.35). The elementary cell (Figure 4.36) is a derivation of the so-called "simplex": The bottom triangle is modified in order to implement contiguous cells. Kono himself brought to my attention a drawing in 1997 (Figure 4.37).

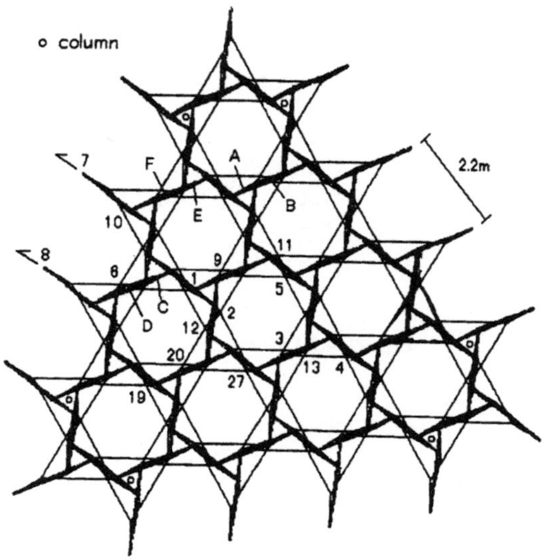

Figure 4.35 *Plane double layer grid by Kono*

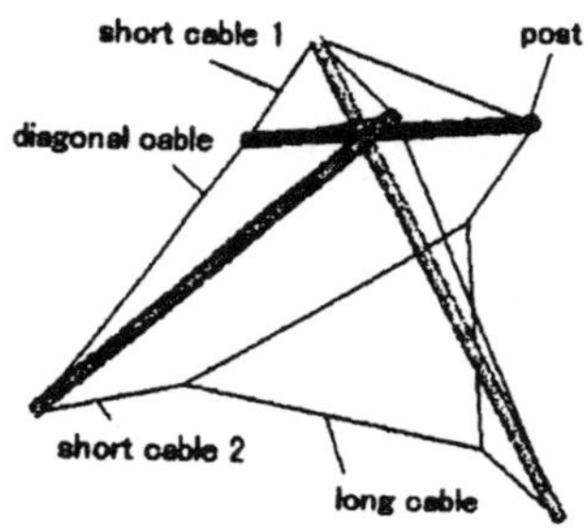

Figure 4.36 *Elementary cell*

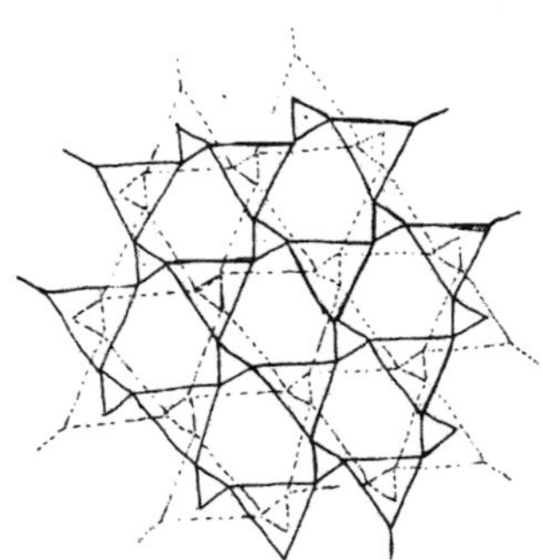

Figure 4.37 *Sketch of cable layers by Kono*

4-431.2 Node on node solutions

Building upon some studies on elementary cells developed by our research team, we decided to work on four strut cells. Three double grids were built with nine elementary cells according to the pattern in Figure 4.38. The following points are worth bearing in mind:

- We tried to respect in this project a rule developed by Robert Le Ricolais by using continuous cables that go through nodes. This is true for layer cables. Bracing cables also constitute continuous paths but it is not at all easy in practical terms to create the whole in one piece because of angles that have to be respected.
- The initial choice to assemble cells by a node-on-node junction could be a reason to exclude these grids from the tensegrity class. B.B. Wang wanted to include them when he wrote about contiguous and non-contiguous tensegrity systems. We think that these grids fit with the extended definition submitted in Chapter 2, in the same way as the tensegrity masts of Snelson and Fuller. It is possible to identify discontinuous sets of compressed components in this type of double layer grid (see Chapter 2).

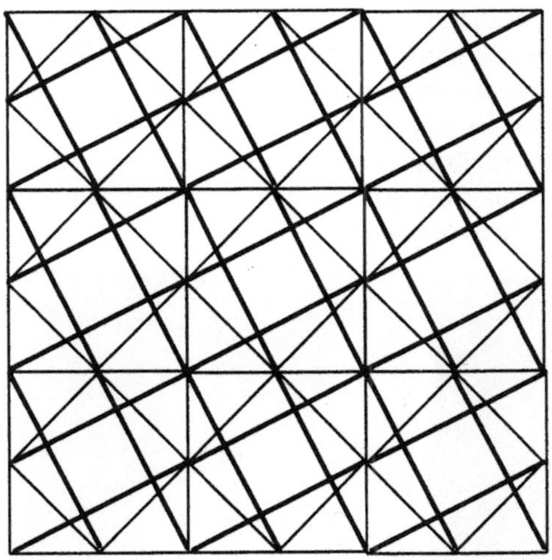

Figure 4.38 *Double layer tensegrity grid principle: plane view*

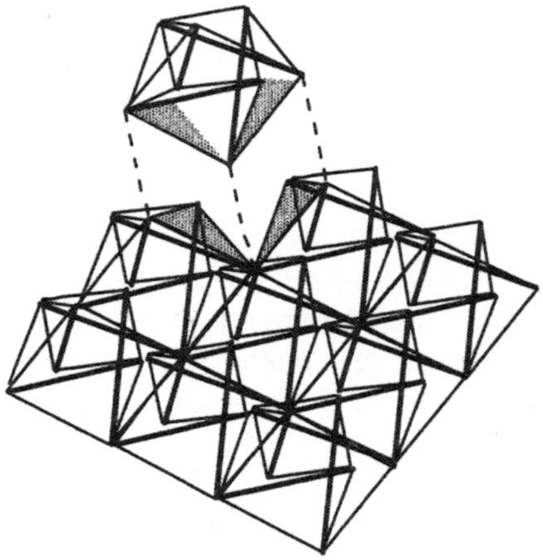

Figure 4.39 *Double layer tensegrity grid principle: axonometric view*

We created three prototypes of this type between 1987 and 1991. The first sketches were produced during the first conference LSA86'[11], but were not published [Ref 4-19]. The first publication of the principle was published in 1987 [Ref 4-20]. These creations required specific studies for nodes. The complete history of these nodes can be found in the work carried out by V. Raducanu [Ref 4-14]. The first attempt was made with aluminium struts and plastic nodes, which were commonly used to display foldable panels by a local firm, named "DUO". We chose to call it a DUO plane double layer tensegrity grid (Figure 4.40). We used the same nodes and constitutive components for grids with curvatures. Curvature results from the variation of distances between nodes.

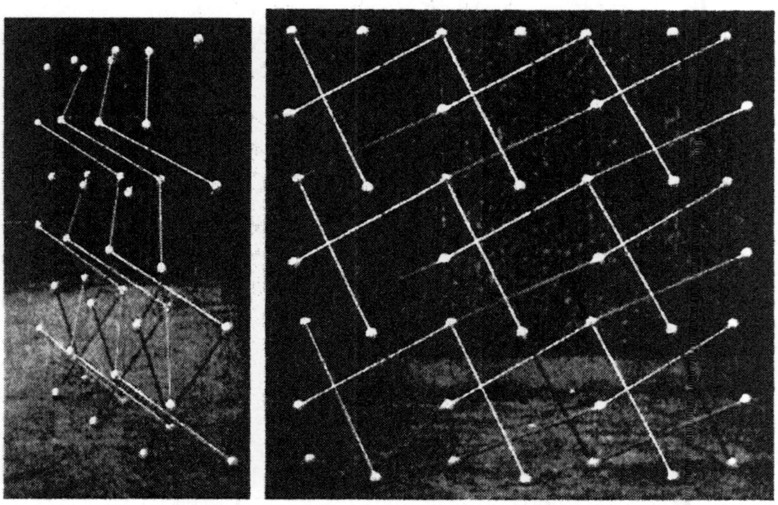

Figure 4.40 *DUO plane double layer tensegrity grid (1990)*

[11] I drew these projects during the conference organised by V. Sedlak in Australia and had the opportunity to introduce the corresponding handmade sketches during my lecture.

A second grid is worthy of mention (Figure 4.41). It was assembled in the same years, but this solution was not developed further. The node had too many degrees of freedom and there was an element of instability which was difficult to overcome. The self-stress level could be introduced by acting directly on the screws at the ends of the struts. Cables go through the nodes that have a cross shape (Figure 4.42).

Figure 4.41 *Double layer plane tensegrity grid*

Figure 4.42 *Node detail*

The third plane double layer tensegrity grid was built at the beginning of 1992, still on the same pattern (Figure 4.43); this grid resulted from our previous studies and we improved the self-stress implementation with the help of my colleague Jean Tuset. Normally a specific device is introduced in struts or cables (a mixed procedure can also be used). In this case we introduced the self-stress with a specific design of the node (Figure 4.44), which contains three parts (Figure 4.45): when assembling these parts bracing cables are tightened and the whole system is under self-stress.

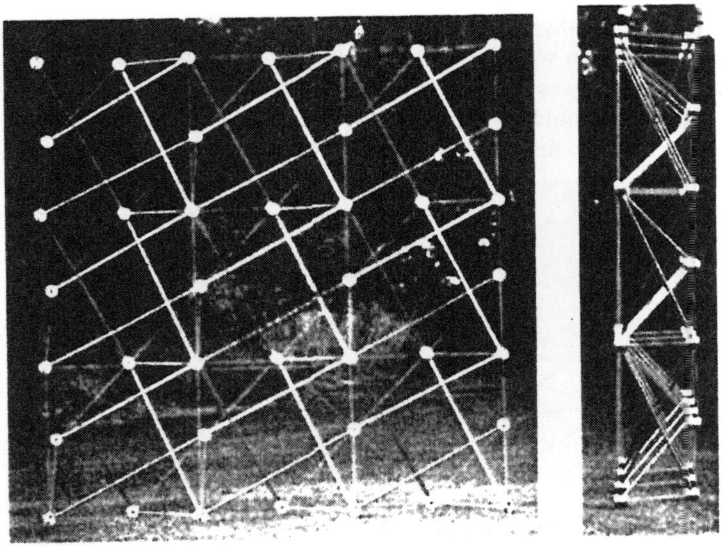

Figure 4.43 *Double layer plane tensegrity grid (1992)*

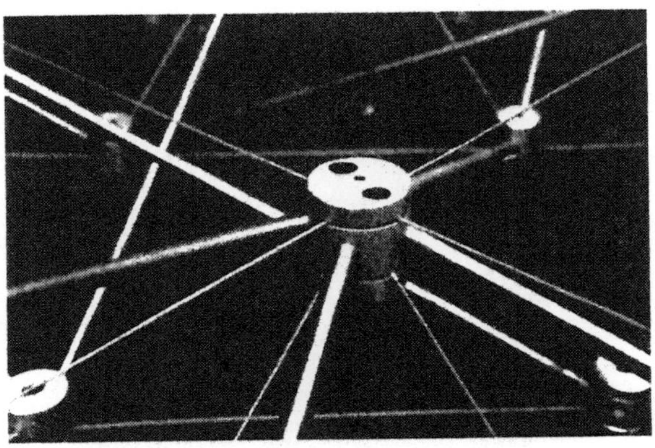

Figure 4.44 *Node in place*

Figure 4.45 *Node parts*

4-43.2. Single curvature systems

One of the goals that has also to be reached is to create single and double curvatures using the tensegrity principle and which does not require any supplementary device such as a mast and/or stabilising cables like for membranes.

By maintaining the principle of elementary self-stressed cells it is possible to modify the equilibrium shape so as to generate single curvature systems. The entire description has been given in [Ref 4-20], and the main features are described here. It is simply necessary to play on the position of nodes referred to as 6 and 8 in Figure 4.46. The design is made with a constant length of the strut; the trajectory of node 6 is a circle in the vertical plane quoted "V".

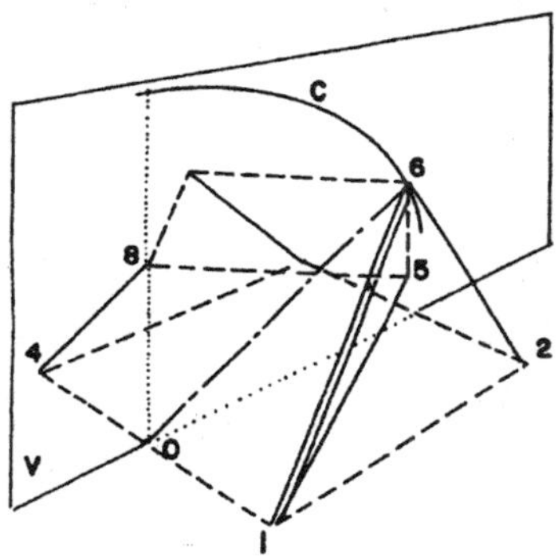

Figure 4.46 *Circular trajectory of node 6*

On this circle the limit positions are defined by the possible strut-strut contacts (Figure 4.47). According to the choice that is operated for the shape modification, two types can be defined (Figures 4.48 and 4.49). We chose type I to design the single curvature double layer tensegrity grid in Figure 4.50.

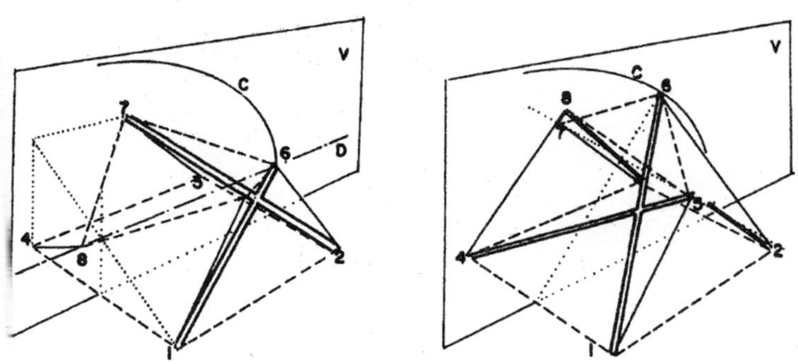

Figure 4.47 *Strut-strut contacts limiting the trajectory*

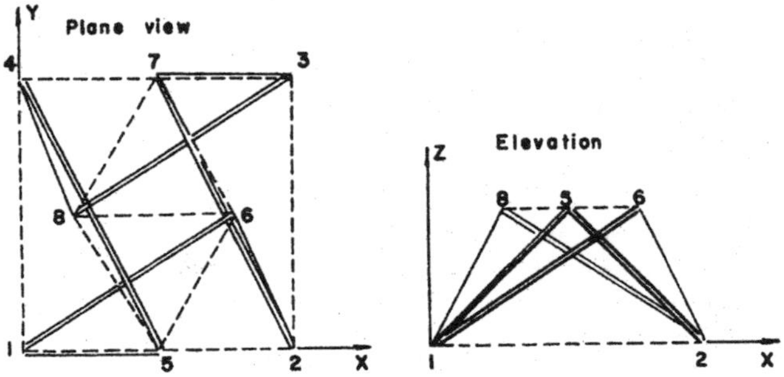

Figure 4.48 *Modified shape type I*

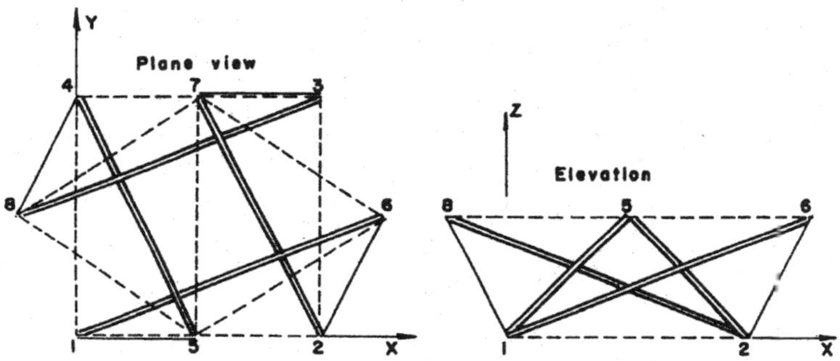

Figure 4.49 *Modified shape type II*

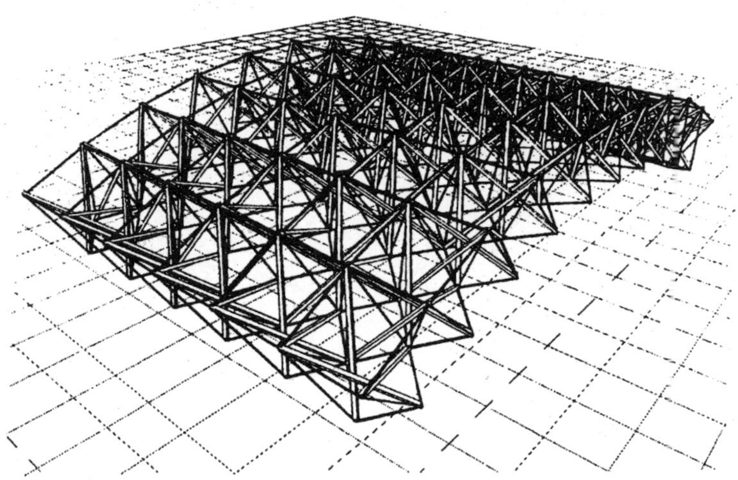

Figure 4.50 *Single curvature double layer tensegrity grid*

4-43.3. Double curvature systems

The final kind of bi-dimensional assembly of cells is a double curvature system. In order to build a prototype of this kind, we chose to generate it as a translation grid. The geometric description can be modelled with Formex algebra [Ref 4-22]: the result is easily achieved by specifying the number of cells on each side, the complete chart requires only a few lines to achieve the result (Table 4.6 and Figure 4.51).

Table 4.6 *Grid generation chart*

```
:C,FL,N,EP:
TC={[1,1,1;3,1,1],[1,1,1;2,1,3],[2,1,3;3,2,3]}
TS=[1,1,1;3,2,3]
FC=RINID(N,N,2,2)|ROSID(2,2)|TC
FS=RINID(N,N,2,2)|ROSID(2,2)|TS
DLG=FC#FS
USE BT(1,1,1)
USE MINE(C,FL,N,EP)
USE VS(10)
USE VN(70,70)
DRAW DLG
```

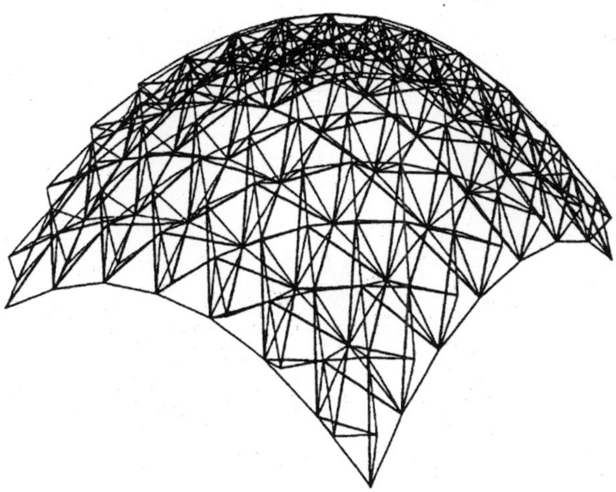

Figure 4.51 *Resulting shape with Formex generation*

The choice was to map cables on two double positive curvature surfaces (a quarter of the whole is represented in Figure 4.56). A project drawing with ten cells on each side was created manually by A. Rampon [Ref 4.23]. These drawings successively represent the two layers of cables (Figure 4.52), the struts (Figure 4.53), struts, bottom layer and bracing cables (Figure 4.54), the grid between the two layers (Figure 4.55) and the final result (Figure 4.56).

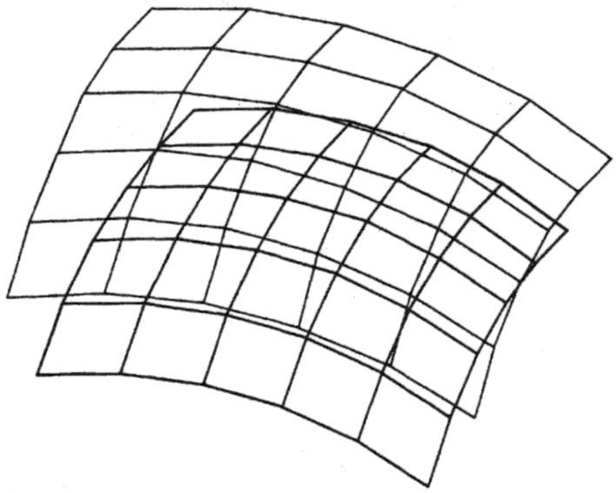

Figure 4.52 *The two cable layers*

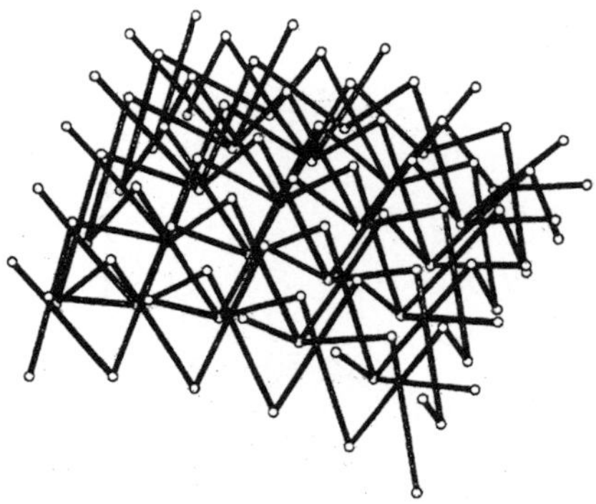

Figure 4.53 *Struts*

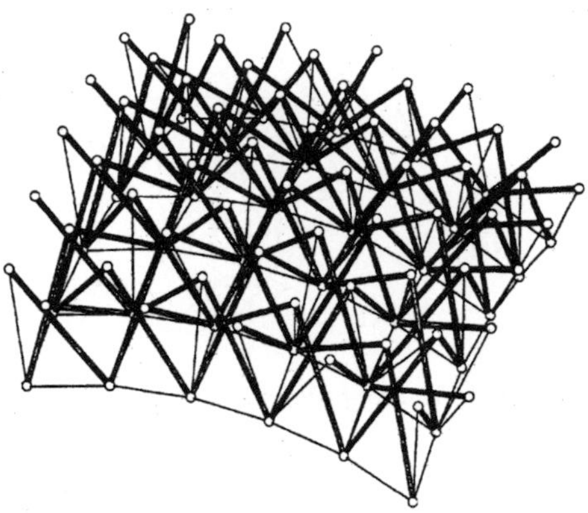

Figure 4.54 *Struts, with bottom layer and bracing cables*

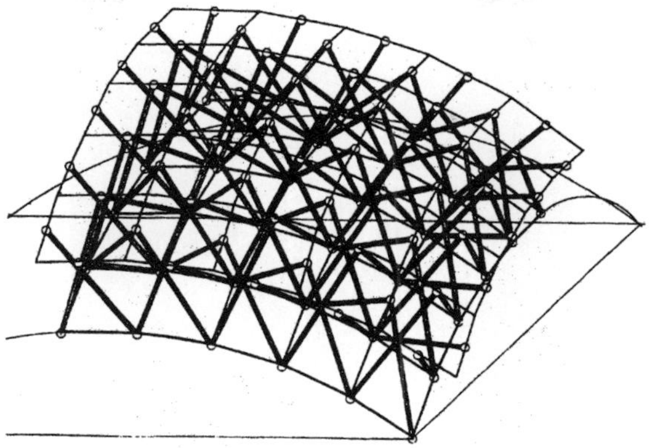

Figure 4.55 *Grid between the two layers*

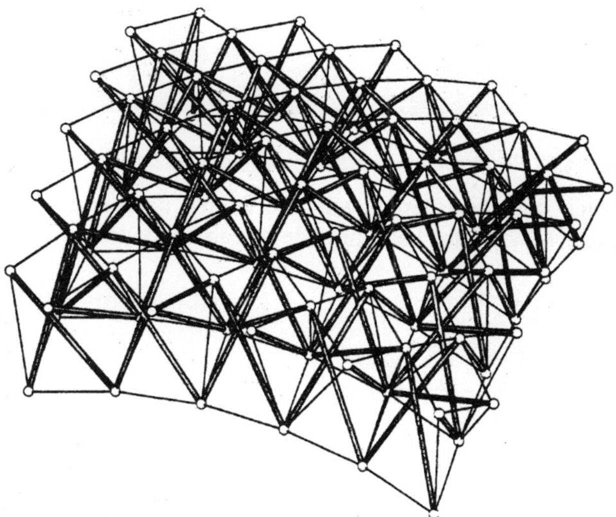

Figure 4.56 *Final shape*

On the basis of these geometrical studies, we created a prototype of a double curvature double layer tensegrity grid (Figure 4.57). We used the same components as those of plane "DUO" double layer grid (Figure 4.40), and this demonstrates that it is possible to vary the shape by playing on the distances between nodes on cables: extra lengths of cables can be seen in the picture.

Figure 4.57 *Prototype of double curvature double layer tensegrity grid*

We will conclude this presentation with the double curvature grid developed by Ariel Hanaor [Ref 4-24] on the basis of node on cable junction.

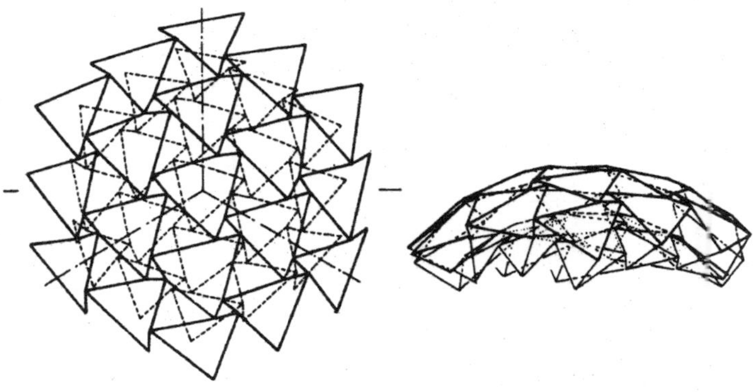

Figure 4.58 *Double layer tensegrity grid by Hanaor*

4-5. Conclusion

In this chapter we have tried to give some examples of typologies based on isolated cells and the assembly of cells. Some are only geometrical studies without equilibrium considerations – which can occasionally lead to incorrect solutions depending on possible realisations; others were effectively created and thus confirming their ability to fit within the concept of structural tensegrity. These typologies are only described from a geometric point of view.

Between them, these studies constitute the history of tensegrity systems, even if some have to be discarded. They opened a wide field which will undoubtedly be investigated by others. The first steps of this new progression will be discussed in the final chapter, for they anticipate a larger development.

References

Ref 4-1. Motro R., Formes et Forces dans les Systèmes Constructifs, Cas des Systèmes Réticulés Spatiaux Autocontraints, Thèse d'Etat, Université Montpellier II, June 1983.

Ref 4-2. Motro R., "Structural Morphology of Tensegrity systems", *International Journal of Space Structures* (Special Issue on Morphology), H. Lalvani Guest Editor, Vol. 11, N° 1 & 2, 1996, pp. 233–240.

Ref 4-3. Emmerich D.G., Cours de Géométrie Constructive – Morphologie, École Nationale Supérieure des Beaux Arts Paris, Centre de diffusion de La Grande Masse, Paris, 1967.

Ref 4-4. Emmerich D.G., Exercices de Géométrie Constructive, École Nationale Supérieure des Beaux Arts Paris, Presses du Centre de Recherches et d'Urbanisme, 1970.

Ref 4-5. Pugh A., An introduction to tensegrity, University of California Press, Berkeley, 1976.

Ref 4-6. Pugh A., Polyhedra, a visual approach, University of California Press, Berkeley, 1976.

Ref 4-7. Minke G., "Tensegrity system", Revue Zodiac, N° 19, 1973.

Ref 4-8. Williamson D., Skelton R.E, "A general class of tensegrity systems: Equilibrium analysis", in Engineering Mechanics for the 21st Century, SCE Conference, La Jolla, 1998.

Ref 4-9. Murakami H., Nishimura Y, "Static and dynamic characterisation of regular truncated icosahedral and dodecahedral tensegrity modules", to appear in *International Journal of Solids and Structures*, 2001.

Ref 4-10. Kenner H., Geodesic maths and how to use it, University of California Press, Berkeley, 1976.

Ref 4-11. Nooshin H., Disney P.L., Champion O.C., "Computer-Aided Processing of Polyhedric Configurations", in Beyond the Cube: The Architecture of Space Frames and Polyhedra, edited by J. François Gabriel, ISBN 0-471-12261-0, Wiley & Sons Inc., 1997, pp. 281–299.

Ref 4-12. Fuller R.B., Marks R., The Dymaxion World of Buckminster Fuller, Anchor Books, 1973.

Ref 4-13. Motro R., "Review of the development of geodesic domes", in Analysis, Design and Construction of Braced Domes, edited by Z.S. Makowski, Granada Publishing Ltd, 1984, pp. 387–411.

Ref 4-14. Raducanu V., Assemblages: conception d'un nœud industriel pour les grilles de tenségrité, Diplôme d'Etudes Approfondies, Université Montpellier II, 1997.

Ref 4-15. Boulet P., "Mât Chinois", Exercices de Géométrie Constructive, École Nationale Supérieure des Beaux Arts Paris, Presses du Centre de Recherches et d'Urbanisme, 1970, p. 93.

Ref 4-16. Chassagnoux A., "Mât Autotendant", Exercices de Géométrie Constructive, École Nationale Supérieure des Beaux Arts Paris, Presses du Centre de Recherches et d'Urbanisme, 1970, p. 211.

Ref 4-17. Hanaor A., "Aspects of Design of Double-layer Tensegrity Domes", *International Journal of Space Structures*, Vol. 7, N° 2, 1992, pp. 101–114.

Ref 4-18. Kono Y., Choong K.K., Shimada T., Kunieda H., "An experimental investigation of a type of double layer tensegrity grids" , *Journal of IASS*, Vol. 41, N° 131, 2000.

Ref 4-19. Motro R., Najari S., Jouanna P., "Tensegrity systems from Design to Realisation", Proceedings of the First International Conference on Lightweight Structures in Architecture, Sydney, Australia, edited by V. Sedlak, 1986, pp. 690–697.

Ref 4-20. Motro R., "Tensegrity systems for Double Layer Space Structures", Proceedings of the International Conference on the Design and Construction of Non Conventional Structures, edited by B.H.V. Topping, Londres, 1987, pp. 43–51.

Ref 4-21. Motro R., "Tensegrity Systems and Geodesic Domes", Special Issue of the *International Journal of Space Structures*, "Geodesic Space Structures", Vol. 5, N° 3, 1990, pp. 343–354.

Ref 4-22. Nooshin H., Formex configuration processing in structural engineering, Elsevier Science Publishers, ISBN 0-85334-315-2, 1984.

Ref 4-23. Rampon A., Formes et structures, Dossier personnel d'études, Ecole d'architecture Languedoc Roussillon, 1988.

Ref 4-24. Hanaor A., "Geometrically Rigid Double-layer Tensegrity Grids", *International Journal of Space Structures*, Vol. 9, N° 4, 1994, pp. 227–238.

5

Models

5-1 Introduction

Some theoretical models were initially developed for form-finding. Other models were necessary to solve some problems related to tensegrity mechanics. I thought it might be fruitful to provide some information on this topic, in order to illustrate the evolution of these models and their relevance today, since many others are to be found in the available literature. I have also included a list of references at the end of this chapter. More references are included in the "Bibliography" chapter at the end of this book.

Needless to say few things can be achieved without the help of others. So it goes without saying that I did not work alone on these models, and I am thus indebted to many colleagues (listed in the references accordingly) and it is my pleasure to warmly thank them all at the very outset of this chapter.

5-2 Problems to solve

It is worthwhile highlighting the main features that govern tensegrity systems, and consequently the associated mechanical problems which have to be addressed.

One is essential: it concerns the initial state of the system and its behaviour when it is subjected to external actions.

The initial state of the system is very specific since it is a self-equilibrated state; moreover the rigidity of the tensioned components is unilateral (no rigidity in compression) and the relational structure is very specific: compressed components are inside a continuum of tensioned components. The study of this initial state, the sizing and sensitivity problems and finally mechanical behaviour (static and dynamic) is the objective of what follows.

As with every system with initial stresses, any state can be defined by two sets of parameters: *form* and *force parameters.*

The first set results from the adopted relational structure and from the geometric characteristics of the manufactured components. Form is directly perceptible and measurable – it can be seen.

The second set of parameters depends on the ability to introduce an initial stress state in the system, which guarantees its stiffness. The level of this stress state cannot be seen directly. In place of stress, it is preferable to evoke the necessary deformations, which are applied to the components to build the system. If there were a strict coincidence between geometric distances and manufactured lengths, the assembly would look like a jigsaw: the only problem would be to find the right place for each piece in the construction.

Some problems to solve are listed below, although list is by no means exhaustive:

1. Form-finding problem;
2. Self-stress feasibility (closely related to point 1);
3. Compatibility between self-stress and component stiffness;
4. Identification of mechanisms;
5. Stabilisation of mechanisms;
6. Sizing of components;
7. Mechanical behaviour under external actions;
8. Sensitivity to imperfections, etc.

The first part of this chapter is devoted to the form-finding problem, and is followed by several others, which are concentrate mostly on mechanical aspects in relation to initial self-stress.

5-3 Form-finding

5-3.1 Introduction

Form-finding is a basic problem for tensegrity systems, since both shape and geometry must fulfil certain stability requirements. Designers have to keep in mind that the double goal to reach simultaneously comprises geometry and self-stress. A form-finding method is characterised by a priority devoted to geometry or to mechanics, but no solution can dispense with either of these two aspects. This is very clear in one of the theoretical methods presented with which the designer may simultaneously use the stress and geometric parameter [Ref 5-1].

It can be claimed that two main methods are available. They can be respectively described as a "form controlled" or a "force controlled" method. The former is illustrated by the work carried out by sculptors in general and by Kenneth Snelson in particular. The objective is to develop tensegrity systems without any criteria about the regularity of components and, moreover, regardless of the generalisation potentiality with mechanical characters. Stability is ensured by a heuristic method based on experimentation and a trial-and-error process. This occasionally gave very impressive results.

The second method was developed in order to ensure the mechanical requirements using a theoretically modelled form-finding process. As one can anticipate, these models have to be simultaneously aware of geometry and pre-stressability in order to be successful. This kind of method can give precious results, but at the same time can fail. It must also be emphasised that if, in this last method, the resulting shapes are very regular; they do not have the richness of the heuristic way. Secondly, it appeared that "mechanically" based solutions require very long development times for complex systems. Consequently, a mixed procedure was embarked upon by our research team: a general principle is defined so as to generate a tensegrity system. The results of its application are then tested according to pre-stressability criterion. We chose this way to design some new tensegrity grids, which will be presented in Chapter 7.

In the following sections, I give a number of illustrations of both methods available to reach this twin goal; these presentations are preferably those which have also a historical interest. When looking at the literature, there is no doubt about the future development of improved methods.

5-3.2 Form controlled methods

5-32.1 Snelson's approach

<u>5-321.1 Basic idea</u>

As previously pointed out (see Chapter 2), Snelson worked only with physical models and sculptures. The basic idea is all contained in its "One to the next" sculpture (Figure 5.1) by transforming the link design between two components. Each has to be considered as an "X" spatial shape, since the lower longer "legs" are used to impose downward forces with clay balls.

The rhombus of cables is then used to link two successive "X". The upper part and the lower part of these "X" are identical in terms of actions: under the effect of load nodes 1 and 2 are discarded and conversely nodes 3 and 4 come closer.

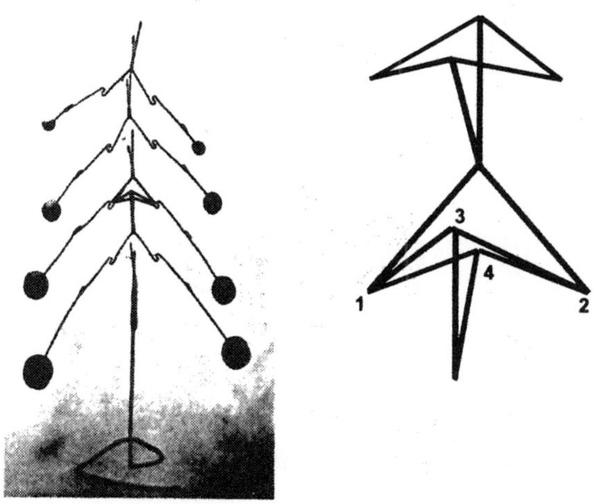

Figure 5.1 *One to the next*

The assembly of "X" naturally leads to the double "X". This evolution is illustrated in Figure 5.2:

- The initial geometry of the "X" is spatial and can be included in a tetrahedron.
- A relative rotation between the upper and lower parts of the "X" gives a plane configuration known as a Saint Andrew Cross (the cross is a compressed component), square edges are tensioned components.
- Using the last configuration and doubling it in an upside down position results in the very well known "double X" sculpture created by Snelson (Figure 5.3).

Figure 5.2 *From tetrahedron to Saint Andrew Cross*

It is worth noting that the junction between the two "X"s is created by a rhombus of tensioned components. When self-weight is considered this rhombus cannot be plane. Consequently, as the size of one of its diagonals increases, so the other decreases: these size variations can be associated respectively with compression and traction when the related actions on nodes like 1 and 2 are considered. This remark is fundamental for the understanding of successive items in the Snelson's patent.

Figure 5.3 *Double "X" sculpture by K. Snelson*

5-321.2 From double "X" to tensegrity

The next steps in the conceptual design achieved by Snelson are described below:

- The initial "X" is split in two pieces, namely "11" and "12" in Figure 5.4.

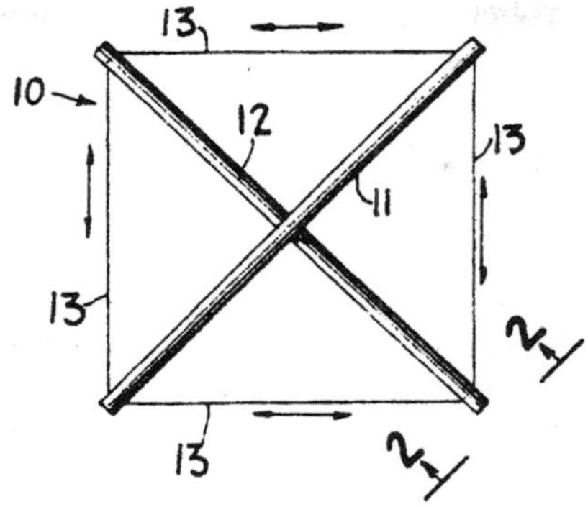

Figure 5.4 *Basic cross*[1]

- Two pieces are assembled: they lie in two perpendicular planes. As previously described (Figure 5.2), one of the four edges is omitted (dashed lines "23" in Figure 5.5), and the link between the two "X" is made with a rhombus of cables ("25" and "24").

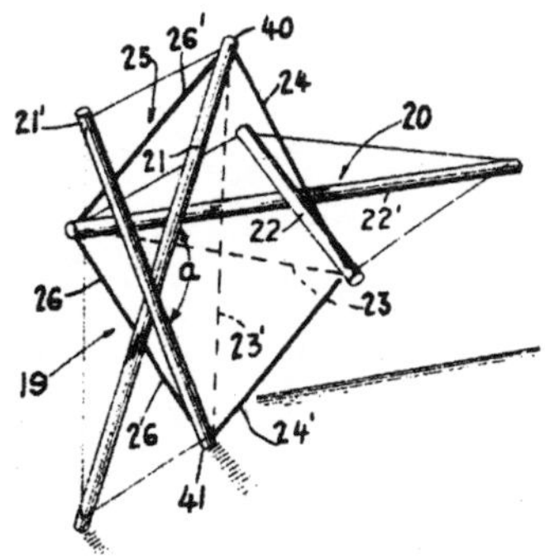

Figure 5.5 *Assembling two "X"*

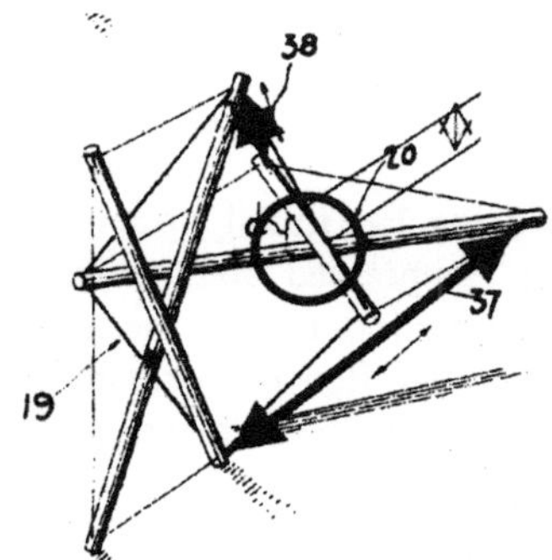

Figure 5.6 *Introduction of two cables*

[1] See Snelson's patent, Chapter 2.

- The addition of tensioned components (red lines) between two "X" allows the separation of the two pieces: their relative distance is illustrated in Figure 5.6. A similar operation on the second "X" gives a system with struts that do not touch each other – thus fitting with the classical definition of tensegrity.

5-321.3 From double "X" to simplex

In paragraph 5-32.1 particular emphasis was paid to the analogy between compression and the increasing size of a distance between nodes 1 and 2. It seems that Snelson used this property himself for in his patent a drawing shows an assembly of three "X". The initial rhombus with four edges becomes a continuous set of nine cables that can be also stabilised by the insertion of struts according to the arrows in Figure 5.7 (right drawing). If these struts are indeed inserted, the system with nine cables and three struts is stable and the three "X" can be removed. Until now it has not been clear to me whether or not if Snelson used this way.

Did Snelson know about Loganson's sculpture and tried to stabilise it? Nevertheless, it struck me as interesting to describe this drawing which is included in Snelson's patent.

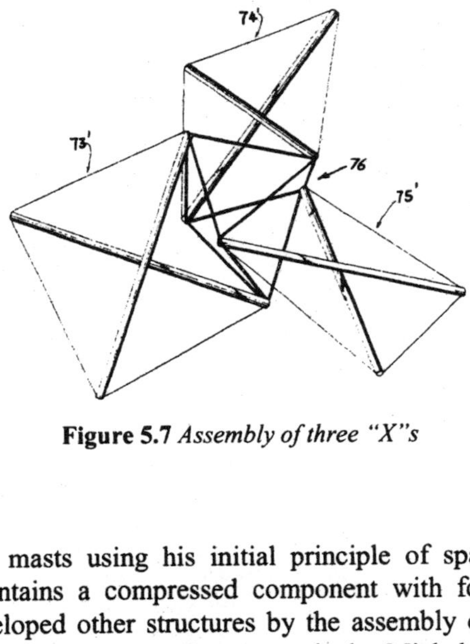

Figure 5.7 *Assembly of three "X"s*

5-321.4 Masts

Snelson built many masts using his initial principle of spatial "X". This basic tetrahedral piece contains a compressed component with four branches and six cables. He also developed other structures by the assembly of several plane "X"s which are separated into two struts. Recent works by Micheletti [Ref 5-2], Skelton [Ref 5-3] and Smaili [Ref 5-4] can be consulted to get a better understanding of the mast design.

Figure 5.8 *Mast (Snelson) and elementary tetrahedral cell*

5-321.5 Conclusion

Snelson's work is fundamental. His design process ensures the pre-stressability of the resulting structures, whose shapes are adapted to the author's aim. But as in every work of this kind, it is not possible to extrapolate generalisations from the process with a series of simple and reliable rules. Apart from Snelson's own experience, it is not possible for anybody to have precise information about the initial state of stress of the whole system during its construction. Moreover, only a great deal of experience may help to monitor this self-stress and its level.

5-32.2 Emmerich's proposals

David Georges Emmerich mainly used a geometric approach, even if he had some contacts with people who carried out mechanical calculations (Siestrunk and Chassagnoux being main ones). He also worked with large scale models which were exhibited in many places.

Nevertheless, each of his proposals needs to be verified in terms of equilibrium. Recent work was undertaken by O. Foucher [Ref 5-5]. The objective was to identify the existence of self-stress states in Emmerich's proposals. Since Emmerich mainly used polyhedra we decided to call those systems "tensypolyhedra" which fit the two mains conditions:

- The nodes are the apices of polyhedra.
- The system has at least one stable self-stress .

<u>5-322.1 Incorrect examples</u>

Many of the drawings made by D.G. Emmerich show polyhedra with struts inside, but not all are "tensypolyhedra". Physical models, numerical models, (mainly those included in Tenségrité 2000[2]), have been used to check the existence of self-stress states. Sometimes simple remarks on equilibrium conditions can lead to rapid conclusions.

a) Truncated tetrahedron

D. G. Emmerich was himself perfectly well aware that the equilibrium configuration is not mapped exactly onto the geometry of the truncated tetrahedron. The hexagonal face is distorted. A simple proof of this impossible equilibrium is illustrated by the projected view given in Figure 5.10: the left-hand side configuration corresponds to the tetrahedron geometry. The equilibrium of node "A" cannot be ensured since the strut does not satisfy the symmetry condition illustrated on the right hand side view. O. Foucher established that it was possible to create a "tensypolyhedron" by the insertion of supplementary cables, inside the "spherical" continuum of cables.

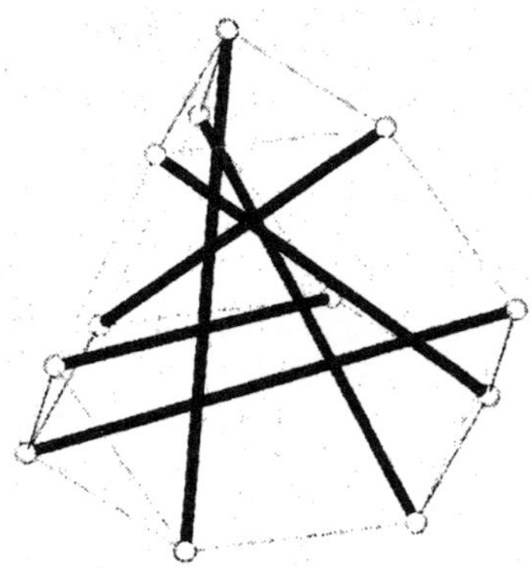

Figure 5.9 *Truncated tetrahedron*

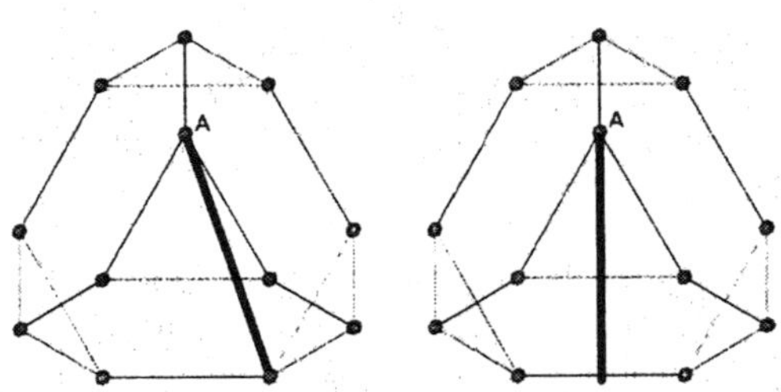

Figure 5.10 *Equilibrium condition*

---

[2] Tenségrité 2000 is a software developed in our own laboratory, dedicated to tensegrity.

b) Icosahedron
Another classical system is based on the geometry of icosahedron with some missing edges between apices whose distance is less than the edge size. This was also pointed by David Georges Emmerich. I established a simple demonstration of equilibrium condition (see paragraph 5-332.2): the ratio between strut length and least distance between two struts is equal to 2. As a matter of proof, it is known that the corresponding ratio in the icosahedron is equal to golden ratio 1,618.

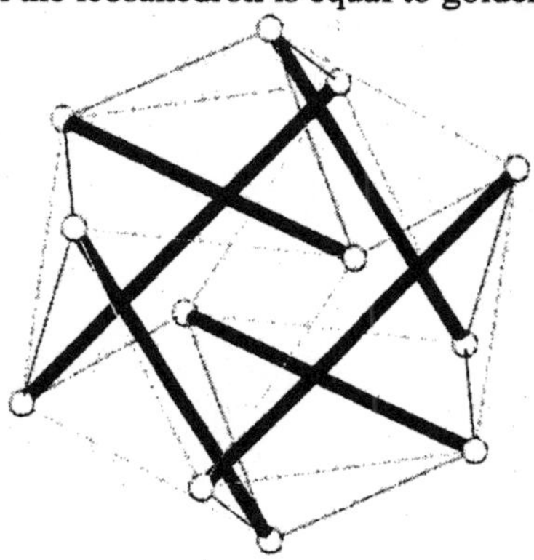

Figure 5.11 *Six strut tensegrity system*

This is why I arbitrarily called it "expanded octahedron" – in order to dissociate this equilibrated configuration from Icosahedron, and also because this configuration could result from an expansion of octahedron: in this geometrical transformation the number of edges (12) is doubled. The three diagonals are split into three pairs of parallel struts.

c) Dodecahedron
This case is similar to the previous one since it contains five pairs of parallel struts (Figure 5 12). A simple physical model shows that this system is subjected to a finite mechanism which can be activated as shown in Figure 5.13. The plane result is also illustrated in the same figure.

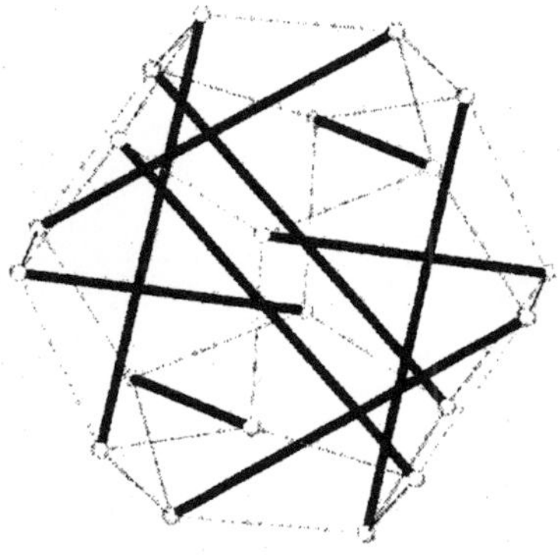

Figure 5.12 *Dodecahedron*

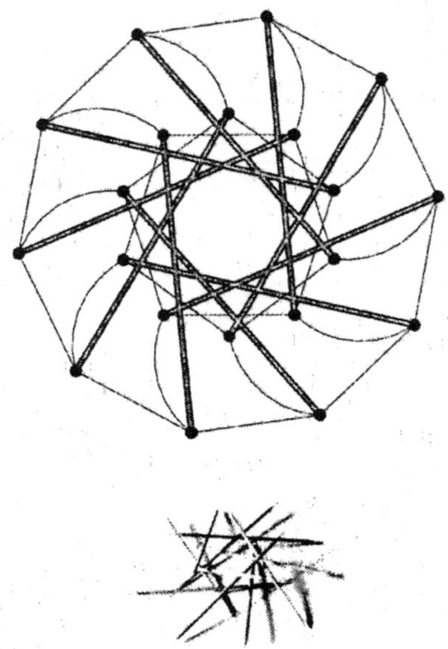

Figure 5.13 *Mapping the dodecahedron on to a plane*

5-322.2 Tensypolyhedra

Several polyhedra can be transformed into tensypolyhedra by insertion of struts. We are only citing here two examples. Interested readers may refer to Foucher's work [Ref 5-5].

a) Snub cube

Among the tensypolyhedra, it is interesting to note the snub cube illustrated in Figure 5.14: it contains 60 cables, 12 struts and 7 self-stress states.

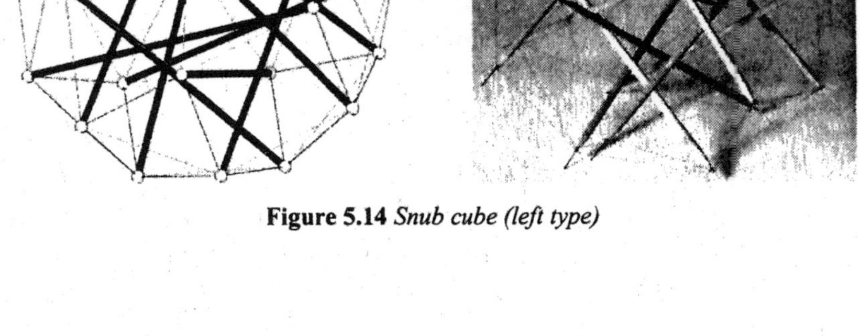

Figure 5.14 *Snub cube (left type)*

b) "Stella octangula"
If we consider the combination of two tetrahedra, we may see it as a starry triangular antiprism. The construction of the tensypolyhedron begins by removing the upper and lower little tetrahedra (Figure 5.15).

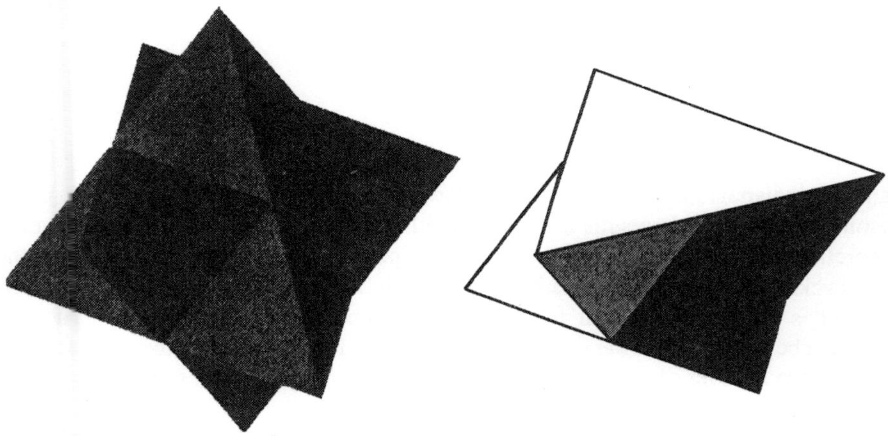

Figure 5.15 *Stella octangula and associate polyhedral shape*

It is then necessary to eliminate some edges so as to keep only those given by Emmerich and reproduced on the right side view (Figure 5.16). Taking advantage of symmetry, we only need to study the equilibrium of a single strut. It can be seen that such a strut (12) is included in a plane rhombus, with one missing edge, which is replaced by the cables incident to nodes 2 and 4. This missing edge is an edge of the original antiprism

Two comments are called for:

- Only two cables are incident to node 1; they define a plane containing the strut.
- This result is easily generalised to other antiprisms with regular polygonal bases.

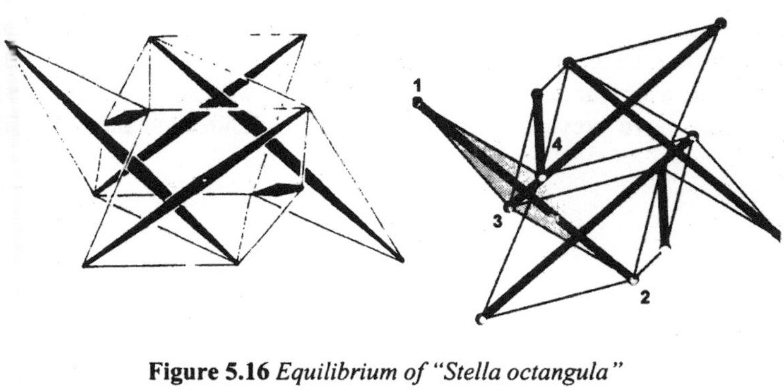

Figure 5.16 *Equilibrium of "Stella octangula"*

5-322.3 Conclusion

The work carried out by Emmerich was very fruitful; in France, many people began to take an interest in tensegrity by reading his publications and/or attending his lectures. I had access to his publications, but I always regretted the absence of stability proofs. This work can now be carried out with theoretical models that are being developed here and there.

5-3.3 Force controlled models

5-33.1 Coupling forms and forces

It is an indisputable fact that Snelson's work were fundamental. And that many results of Emmerich's results are very useful. But it was also necessary to develop theoretical models that could be used by people who want to overcome the drawbacks of experimental methods based on trial and error processes: if this method is very well adapted for art, it is not necessarily the case for structural design. The other side of the coin is, of course, that it is difficult to take into account a very large number of parameters when mechanical methods are used.

The goal is to design tensegrity systems ensuring the coupling of forms and pre-stressing forces in the final result, as a double simultaneous target. Analytical methods can be developed for cases with a low number of parameters [Ref 5-6]; numerical methods allow for a larger number. It is obvious that each model has its own limits and the designer might be well advised to use several so as to improve his own understanding.

5-33.2 Form-finding: single parameter

Apart from sculptural works, until now most tensegrity systems have been regular, and their shape is defined by only one parameter, the ratio r = s/c – s being the length of the struts and c the length of the cables. In this case form-finding is a *single parameter* process.

5-332.1 Static equilibrium approach

For very simple systems like the three-strut tensegrity module, a static approach can be used. The equilibrium conditions for one node determine the resulting shape. This demonstration has been given in Chapter 3 devoted to "Fundamental Concepts". Here we operate a similar demonstration for a system with six struts and twenty-four cables, that we have previously referred to as "expanded octahedron" [Ref 5-7] Figure 5.17.

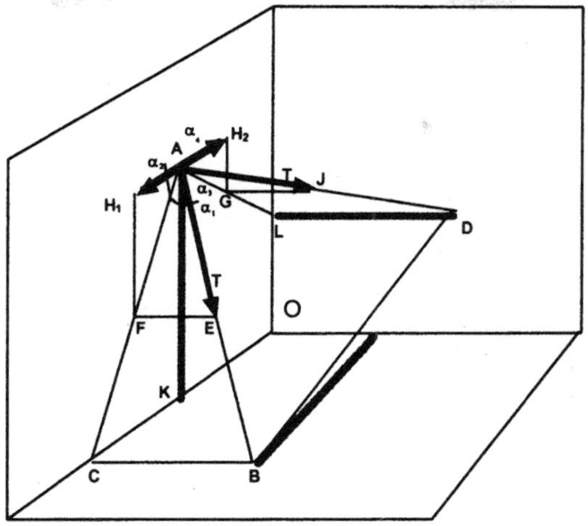

Figure 5.17 *Expanded octahedron. Node equilibrium and form-finding*

For reasons of symmetry, it is sufficient to consider only 1/8° of the whole system, and to study the equilibrium of one node; let "A" be this node. Five actions are exerted on it, one by the strut and four by the four cables. Let "T" be the action of one of these cables. Again for reasons of symmetry, we may study only two, which are on one side of the plane containing A; their components can be split into components that are orthogonal to the vertical component containing the strut: these components are self-balanced when the four cables are considered. The in plane components can be split into vertical components equilibrated by the compression in the strut and two horizontal components, say H_1 and H_2. Equilibrium condition is given by:

E 5.1 $$H_1 = H_2$$

with:

E 5.2 $$H_1 = T.\cos\alpha_1 \cdot \cos\alpha_2$$

and

E 5.3 $$H_2 = T \cdot \cos\alpha_3 \cdot \cos\alpha_4$$

It can be seen in Figure 5.17 that trigonometric lines are given by following values:

E 5.4 $$\cos\alpha_1 = \frac{AC}{AB} \qquad \cos\alpha_2 = \frac{CK}{AC}$$

and then:

E 5.5 $$H_1 = \frac{CK}{AB} \cdot T$$

Similarly

E 5.6 $$\cos\alpha_3 = \frac{AL}{AD} \qquad \cos\alpha_4 = \frac{KO}{AL}$$

E 5.7 $$H_2 = \frac{KO}{AD}\cdot T$$

Equilibrium equation E 5.1 becomes:

E 5.8 $$\frac{CK}{AB} = \frac{KO}{AD}$$

The system is regular and then:

E 5.9 $$AB = AD = c$$

with c being the cable length. Consequently:

E 5.10 $$CK = KO \Rightarrow \frac{s}{4} = \frac{d}{2}$$

s being the strut length, and d the distance between two struts. The equilibrium position is then characterised by the very simple relationship:

E 5.11 $$\mathbf{s = 2\,d}$$

It can be demonstrated easily that the ratio between the strut length and the cable length is:

E 5.12 $$\frac{s}{c} \approx 1{,}63$$

This self-stress state is also characterised by the relationship between the compression C in the struts and the traction T in the cables:

E 5.13 $$C \approx 2{,}45T$$

5-332.2 Cinematic approach

This approach is illustrated for the three-strut cell.

If we express the ratio r = s/c with respect to θ, which is the relative rotation between the two parallel equilateral triangles (case of regular three strut module), we find that:

E 5.14 $$r = \frac{s}{c} = \left[1 + \frac{2}{\sqrt{3}}\cdot\sin\left(\theta + \frac{\pi}{3}\right)\right]^{\frac{1}{2}}$$

The null self-stress equilibrium geometry is defined by the maximum value of the ratio r; hence we derive the expression E 5.14:

E 5.15 $$\frac{d\left(s/c\right)}{d\theta} = \frac{\sqrt{3}}{3} \cdot \cos\left(\theta + \frac{\pi}{3}\right) \cdot \left[1 + \frac{2}{\sqrt{3}} \cdot \sin\left(\theta + \frac{\pi}{3}\right)\right]^{-\frac{1}{2}}$$

and find two values of θ:

E 5.16 $$\theta = \pm\frac{\pi}{6}$$

Related systems are called "anti-clockwise" ("lévogyre"), and "clockwise" ("dextrogyre") (Figure 5.18). It is possible to check, in these solutions, the parallelism of lines previously called D and D', in vertical projection.

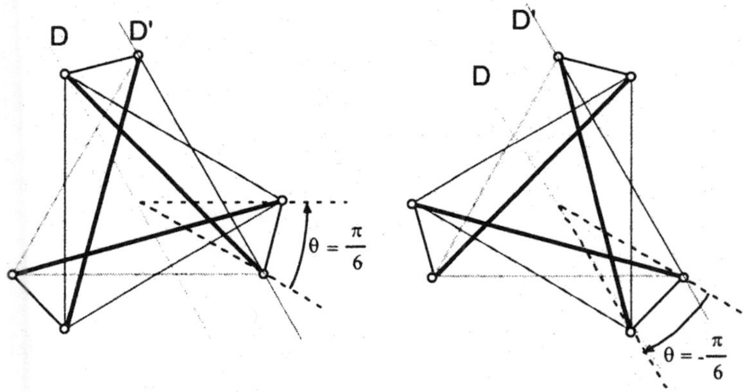

Figure 5.18 *"Anti-clockwise" and "clockwise" systems*

It is possible to generalise [Ref 5-8] the equation E 5.14 in the case of p-polygonal systems. I have demonstrated that it leads to the following expression:

E 5.17 $$\theta = \pm\frac{\pi \cdot (p-2)}{2 \cdot p}$$

But as Pugh pointed out in his book [Ref 5-13], this relation links the two bases without any information about the depth of the module.

5-332.3 Dynamic relaxation

A third approach can be used for single-parameter systems. This approach is based on dynamic relaxation with, or without, kinetic damping [Ref 5-9]. In this numerical method, static equilibrium is the result of a virtual dynamic study of a system. This approach used for a truncated tetrahedron shows that the hexagonal faces, such as ABCDEF, are not planar but slightly distorted (Figure 5.19).

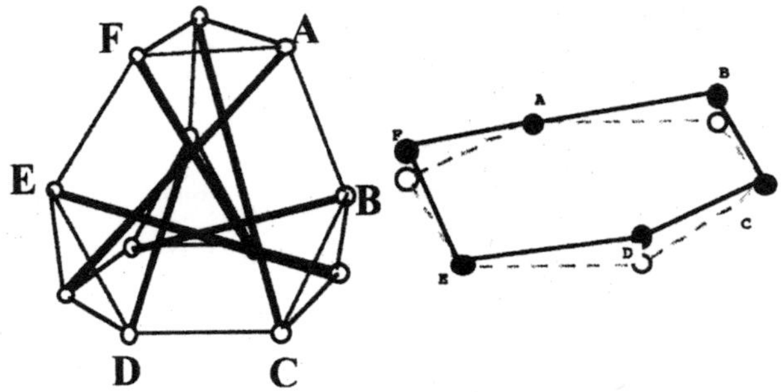

Figure 5.19 *Truncated tetrahedron*

5-33.3 Form-finding: multi-parameter

Multi-parameter form-finding processes have been developed recently for irregular modules and assemblies. This new method has been developed mainly by Vassart [Ref 5-1]; it is based on the force density approach. Irregular shapes can be defined either by choice of force density coefficients or by an analytical method. Graphic developments have also been studied in this work and it is now possible to design irregular shapes and to try to map them on double curvature surfaces (with positive and/or negative gaussian curvature). This is a fundamental step for the form-finding of tensegrity systems. The main features of this method are outlined in this section.

5-333.1 Force density method

Force density principle.

The force density method [Ref 5-1] is based on the force density coefficients {q}, which are used to linearise the equilibrium equations. For a member j, the force density coefficient q_j is defined by the ratio between the normal axial force T_j and reference length l_j^0:

E 5.18 $$q_j = \frac{T_j}{l_j^0}$$

Reference length l_j^0 for a member j represents its length after assembly of the whole system and before any loading. Let "i" be a free node, connected with nodes "h" by members j, equilibrium equation for "i" along X direction s:

E 5.19 $$\sum_j \frac{(x_i - x_h)}{l_j^0} \cdot T_j = f_{ix}$$

And consequently with introduction of force density coefficients:

E 5.20 $$\sum_j (x_i - x_h) \cdot q_j = f_{ix}$$

x_i is the coordinate of node i in X-direction, and f_{ix} the component of the external load $\{f_i\}$ applied at "i" along the X-direction. $\{f_i\}=0$ for form-finding.

Matrix form of equilibrium with force density method.
In the case of a reticulate system with b members and n nodes, the preceding equilibrium equation can be written in matrix form as:

E 5.21 $$[C]_i^t \cdot [Q] \cdot [C] \cdot \{x\} = f_{ix}$$

where [C] is the b x n connection matrix of the studied reticulate system, $[C]_{,i}^t$ is the ith column vector of the transpose matrix $[C]^t$) and [Q] is the b- square diagonal matrix, comprising the b force density coefficients. [C] is regular and is constructed line by line for each member m whose extremities are k and g (with k<g). All terms of this matrix are equal to 0, excepted:

E 5.22 $$c_{mk} = -1$$

and

E 5.23 $$c_{mg} = 1$$

For the n nodes of a reticulate system, we may write:

E 5.24 $$[C]^t \cdot [Q] \cdot [C] \cdot \{x\} = f_x$$

with vectors $\{x\}$ and $\{f_x\}$, being respectively the x-coordinate of all n nodes and the external actions along the x-direction. Values that are associated with nodes are split into two parts to introduce boundary conditions. This leads to a partition of the connection matrix. The first part is built with the terms related to free nodes (subscript "l", and subscript "l_x" for x-direction), the second part with those fixed in the considered direction (the subscript "f" is used for these values, and "f_x" for x-direction). If, in equations [E 5.24], we consider only the n_{lx} equilibrium equations associated with n_{lx} free nodes, we have:

E 5.25 $$[C_{lx}]^t \cdot [Q] \cdot [C_{lx}] \cdot \{x_l\} = \{f_{lx}\} - [C_{lx}]^t \cdot [Q] \cdot [C_{fx}] \cdot \{x_f\}$$

We define the "connectivity matrix" $[D_x]$ containing the force density coefficients of nodes, which are free along the X- direction, as follows:

E 5.26 $$[D_x] = [C_{lx}]^t \cdot [Q] \cdot [C_{lx}]$$

And, if we note:

E 5.27 $$[D_{fx}] = [C_{lx}]^t \cdot [Q] \cdot [C_{fx}]$$

we obtain the equilibrium equations in matrix form for the force density method, along the X-direction:

E 5.28 $$[D_x] \cdot \{x_l\} = \{f_{lx}\} - [D_{fx}] \cdot \{x_f\}$$

Similar expressions hold in the Y and Z-directions:

E 5.29 $$[D_x] = [D_y] = [D_z]$$

and

E 5.30 $$[D_{fx}] = [D_{fy}] = [D_{fz}]$$

Force density method and form-finding.
In the case of form-finding for pre-stressed or self-stressed[3] reticulate systems, no external actions are considered in the search for equilibrium, except for actions that are related to fixed node for pre-stressed systems . For such cases, the force density coefficients $\{q\}$ are called pre-stress or self-stress coefficients and denoted by $\{q^0\}$. Consequently, the equilibrium equation [E 5.28] becomes:

E 5.31 $$[D_x^0] \cdot \{x_l\} = -[D_{fx}^0] \cdot \{x_f\}$$

With $[D_x^0]$ being a n_{lx}. x n_{lx} connectivity matrix containing pre-stress (or self-stress) coefficients of nodes free in the X-direction:

E 5.32 $$[D_x^0] = [C_{lx}]^t \cdot [Q^0] \cdot [C_{lx}]$$

$[D_{fx}^0]$ is the n_{lx} x n_{fx} connectivity matrix, which is defined by:

E 5.33 $$[D_{fx}^0] = [C_{lx}]^t \cdot [Q^0] \cdot [C_{fx}]$$

In these two last expressions $[Q^0]$ is the diagonal matrix that is formed with the b pre-stress (or self-stress) force density coefficients.

The matrix $[D_x^0]$ is easily determined from the relational structure of the system and the chosen pre-stress (or self-stress) coefficients (see [E 5.32]). When the relational structure of a reticulate system is known, its connection matrix [C] is defined. Moreover, the chosen pre-stress (or self-stress) coefficients determine matrix $[Q^0]$ and then matrices $[D_x^0]$ and $[D_{fx}^0]$ are known.

Finding the unknown coordinate (i.e. $\{x_l\}$,...) associated with the chosen pre-stress (or self-stress) coefficients, is achieved by solving a linear system of equations such as the X-direction equation (i.e. $[D_x^0] \cdot \{x_l\} = -[D_{fx}^0] \cdot \{x_f\}$).

---

[3] Self-stressed systems are a specific case of prestressed systems: initial stresses result only from the assembly. For prestressed systems, at least one other system is associated (like compression ring or masts...)

In these equations, the fixed node coordinate vector $\{x_f\}$ is given in the case of a pre-stressed system, and they are identically null vectors in the case of a self-stressed system.

5-333.2 Application to self-stressed reticulated systems

Equilibrium of self-stressed reticulated systems.
When self-weight is neglected, a self-stressed system does not require any fixed node. The self-stressed geometry is defined by the relative position of the nodes, and the system can be considered as free, forming a rigid body free in space. Whatever the coordinate system OXYZ, the three n x n matrices that define the self-stress coefficient connectivity matrix are identical:

E 5.34 $$[D_x^o] = [D_y^o] = [D_z^o]$$

And in this case the relationship that links these matrices to the connection matrix [C] is:

E 5.35 $$[D_x^o] = [D_y^o] = [D_z^o] = [C]^t \cdot [Q^o] \cdot [C]$$

Three homogeneous equation systems have to be solved; they are identical to the following equation written for the X-direction:

E 5.36 $$[D_x^o] \cdot \{x_l\} = \{0\}$$

Rank of self-stress coefficient connectivity matrix.
Firstly, it is necessary to emphasise the fact that these matrices are always singular, since for any column or row the sum of all terms is always equal to zero (Vassart, [Ref 5-1]).

When there are n free nodes, then:

E 5.37 $$\text{rank}([D_x^o]) = \text{rank}([D_x^o]) = \text{rank}([D_x^o]) < n$$

Consequently, the equilibrium equations admit an infinite number of solutions, since all characteristic determinants vanish when the system to be solved is homogeneous.

In the case of non-specific self-stress coefficients, the rank of these matrices is generally equal to n-1. In this case the solutions can be parameterised in terms of only one redundant coordinate, hence all the other nodes coincide with this redundant node. In order to have self-stressed geometries that are not restricted to one point (or one straight line), it is necessary to reduce the connection matrix rank to n-2 (respectively n-3).

When this rank is equal to n-3, the solutions are parameterised in terms of three redundant coordinates. The resulting self-stressed forms are then plane in the best instances. For plane reticulate systems it is sufficient, but for spatial systems, it is necessary to further reduce the rank by one.

When the connectivity matrix rank is reduced to n-4, the four redundant nodes that parametrise the solutions are then sufficient to generate spatial reticulated self-stressed systems.

Form-finding process for reticulated self-stressed systems.
Carrying out the form-finding of a self-stressed reticulate spatial system, defined by its relational structure, with the force density method, consists of three distinct steps:

- Step 1: Find self-stress coefficient values such that the rank of self-stress coefficient connectivity matrix is less or equal to n-4 for a spatial system (or n-3 for a plane system).
- Step 2: Solve the linear homogeneous system of equilibrium equations written with the chosen coefficients.
- Step 3: Identify the form required among the solutions obtained with the chosen coefficients.

The main difficulty is in the first step, for which several techniques can be used to find solutions; some of which are listed below:

- *Intuitive method:* adopted for systems with only few members; this method is not recommended for other cases.
- *Iterative method:* self-stress coefficients are evaluated step by step until the rank of connectivity matrices reaches the required order. Because of recent improvements in computing efficiency, less time is required, and this method provides interesting results.
- *Analytic method:* the matrices are analysed in their symbolic form in order to find the self-stress coefficients that satisfy the required rank condition. This method can be regarded as optimal. But a semi-symbolic approach (i.e. some self-stress coefficients are chosen) can also be used when the number of members is large.

Of course, these techniques can also serve for pre-stressed reticulate systems when the self-stress coefficient connectivity matrices have to be singular and when all the characteristic determinants are reduced to zero.

5-333.3 Application to tensegrities

Form-finding of triplex.
We name *triplex* every tensegrity system comprising six nodes, three struts and nine cables, so that every node is connected to one strut and three cables. All associated systems are defined by the relational structure of the so-called simplex, which is a regular tensegrity system.

In order to find irregular triplex we know that it is necessary to reduce the rank of the self-stress coefficient connectivity matrix to n-4. The matrices being of dimension 6, their rank will be equal to 2. The required self-stress coefficients have to be different from zero and satisfy the conditions $q^0 > 0$ for cables and $q^0 < 0$ for struts. All self-stress coefficient combinations which lead to a rank equal or less than two are acceptable.

If, for instance, we choose identical coefficients for members pertaining to the same set of members, that is:

- lower triangle cables:

E 5.38 $$q_1^0 = q_2^0 = q_3^0 = q_i^0$$

- bracing cables:

E 5.39 $$q_4^0 = q_5^0 = q_6^0 = q^0$$

- upper triangle cables:

E 5.40 $$q_7^0 = q_8^0 = q_9^0 = q_s^0$$

- struts:

E 5.41 $$q_{10}^0 = q_{11}^0 = q_{12}^0 = -q^0$$

We find after an analytical study (based on Gaussian elimination) that the following relationship has to be satisfied in order to reduce the rank to two:

E 5.42 $$3 \cdot q_i^0 \cdot q_s^0 - \left(q^0\right)^2 = 0$$

Consequently if:

E 5.43 $$q_1^0 = q_2^0 = 1$$

Then q^0 can be derived:

E 5.44 $$q_1^0 = q_2^0 = 1 \Rightarrow q^0 = \sqrt{3}$$

For every direction, there remain only two independent equilibrium equations. There are then four redundant nodes, which can be located anywhere. Let ❶, ❷, ❸ and ❹ be these nodes. The location of the two others, ⑤ and ⑥, is then defined on the basis of the two independent equations. For instance, if ❶, ❷ and ❸ are vertices of an equilateral triangle, and node and ❹ is chosen so that the upper and lower triangles are centred one with respect to the other, the regular triplex is called *simplex*.

With still the same values, triplex with irregular bases can be determined. Figure 5.20 shows the case of a rectangular isosceles triangle for the bottom basis with a resulting isosceles triangle for the top basis.

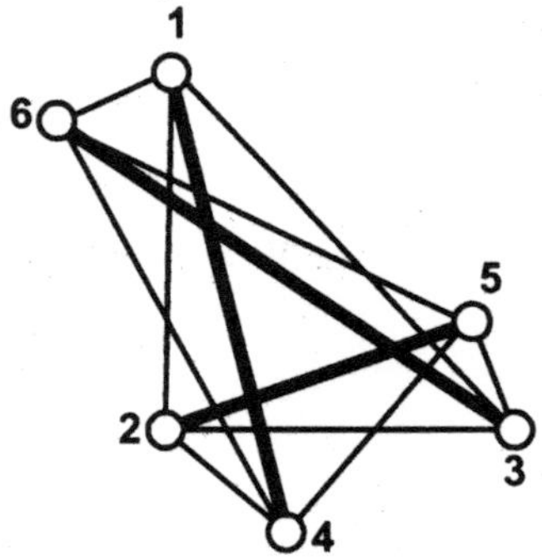

Figure 5.20 *Triplex with isosceles triangles*

A whole family of geometries can be defined with the same set of self-stress coefficients using the redundant nodes (linked thus with form parameters). A new set of self-stress $\{q^0\}$ coefficients will give access to a new family of forms.

For instance we now take:

E 5.45 $$q^0 = q_s^0 = 1$$

we obtain the relation:

E 5.46 $$q_i^0 = \left(q^0\right)^2 / 3 \cdot q_s^0 = 1/3$$

Then, locating ❶, ❷ and ❸ being as before at the vertices of an equilateral triangle, and choosing ❹ such as to centre vertically the two triangles one above the other, we reach a triplex (Figure 5.21). The top vertices in a plane projection are located on the edges of the bottom triangle (we refer to it as an "inscribed top triangle"). This "inscribed top triangle triplex" is interesting since it allows the generation of plane double layer grids when associated with other similar anti symmetric modules.

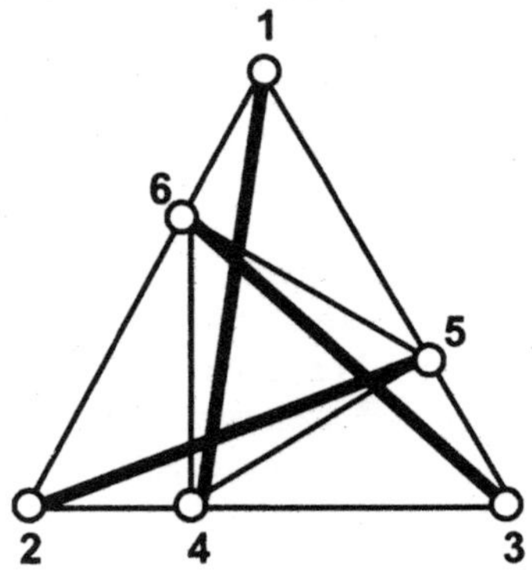

Figure 5.21 *Triplex with "inscribed top triangle"*

5.333.4 Concluding comments

It has been shown that the force density method is well suited to the form-finding of cable nets, which are purely tensioned reticulate systems. By successive modifications of the pre-stress coefficients, several pre-stressed forms, satisfying prescribed boundary conditions may be computed.

For reticulate pre- or self-stressed systems comprising tensile and compressive members, this method offers new perspectives since it allows relatively simple multi-parameterised form-finding processes. Some examples have been shown for the cases of "triplex" and of "quadruplex" structures.

Apart from the relational structure, which is assumed to be known at the beginning of the process, two sets of form-finding parameters can be identified. Pre- or self-stress coefficients constitute the first set, ("force set"). They have to satisfy certain conditions and their choice is by no means easy for the uninitiated; this choice influences the resulting form. At the beginning, this method was used for cable nets and all members were under tension, consequently all coefficients had the same sign. However, for other systems, the designer can prescribe either compression or tension for each member, this stress being associated with the sign of the coefficients. The second set of parameters ("form set") is related to the coordinates of the redundant nodes. No restriction is imposed in this case and the modification of form deriving from this choice is very large.

It can be underlined that the range of pre- or self-stress shapes is directly related to the number of restrictive conditions imposed by the designer.

The form-finding of tensegrity systems is a key stage in their study. If numerical methods like force density are useful for complex systems comprising a great number of elements, it should be kept in mind that sometimes some simple static equilibrium considerations could help avoid misunderstandings.

5-33.4 Form-finding and pre-stressability

Form-finding processes are not independent of pre-stressability. Different approaches have to be underlined. Some methods are built on an assumption of pre-stressability and it is not surprising that the resulting systems can be self-stressed.

The self-stress problem is addressed in the following section and developed accordingly. The main idea to keep in mind is that the resulting self-stress needs to be a feasible self-stress (that prevent cables from compression), and this feasibility is always checked.

Besides theoretical form-finding numerical methods, a number of other paths are opened:

- analytical methods (like in section 5-332.3), which can only be developed for very simple cases. Recent works, mainly by Skelton and Sultan are based on this approach. A comprehensive list of their publications is given in references [Ref 5-14] to [Ref 5-18].

- trial and error process or what could be called "Snelson process", which is based on progressive experiments; the pre-stressability is checked throughout the process. Lengths are adapted to the possibility of self-stress.
- conceptual methods based on a structural composition fitted to tensegrity in unspecified places of the system and combined with suitable boundary composition. This method was successfully used in our laboratory during the "Tensarch" project.

Whatever the chosen method, if the feasibility of the whole self-stress is ensured, designers have to solve a very specific problem related to the implementation of self-stress. This problem had also to be solved for membranes: it is necessary to have the equivalent of the "cutting pattern" in order to assemble components in such a way that at the end the system is simultaneously at the right geometry and at the chosen state of initial stress. This is not an obvious question since many factors have to be considered: sensitivity to manufacturing imperfections, choice of active components, gradual implementation of self-stress, control of the self-stress distribution, and, finally, simultaneous control of the geometry.

5-33.5 Stability of pre-stress

N. Vassart *et al* [Ref 5-17] submitted an energetic stability criterion of the self-stress state. An illustration of stability study is given in the following sections for a very simple case of two aligned components. The description of the associated method is outside of the scope of the present work. But a very significant result can be drawn on this basis. If we consider a reverse system of the so-called "simplex" containing nine compressed components and three tensioned components, it can be established that the corresponding state of self-stress is not stable.

5-3.4 Conclusion

This section was written in order to give an illustration of the first theoretical and practical methods used for form-finding approaches. We underlined the force density method, because of its generality and its ability to very easily provide results for irregular systems. These are very difficult to study with analytical models which are themselves successful when a very small number of parameters is concerned. It appears that these two theoretical models, one numerical and the other analytical are actually improved, but our presentation has to be considered only as an historical review.

5-4 Self-stress and mechanisms

5-4.1 Introduction

Apart from form-finding another class of problem is characteristic of tensegrity systems i.e. rigidity: this class is mainly concerned with self-stress. Most tensegrity systems have infinitesimal mechanisms, and may, or may not, be stabilised by the self-stress state(s). The stability of the self-stress state previously described for the four-strut tensegrity system is ensured: if one or more nodes are displaced from their initial location by an appropriate action which activates an infinitesimal mechanism, they will return to their original location if this action is suppressed. If this self-stress state stabilises the infinitesimal mechanism, it is not the case when struts and cables are exchanged (which would correspond to the case of a dual module with four cables and twelve struts). Problems that are related to self-stress states, infinitesimal mechanisms and stability are treated in the following sections.

5-4.2 Tensegrity systems and reticulate systems

Tensegrity systems are a subclass of reticulate systems, and consequently, it is useful to describe the theoretical basis, which is necessary to understand problems that are related to self-stress and mechanisms.

Generally for a spatial reticulate system with b members with bilateral rigidity, and N degrees of freedom, the number "ss" of self-stress states is given by:

E 5.47 $$ss = b - r_A$$

and the number "m" of internal mechanisms by:

E 5.48 $$m = N - r_A$$

and consequently:

E 5.49 $$ss - b = m - N$$

Where r_A is the rank of the equilibrium matrix [A], which verifies the following equilibrium equation:

E 5.50 $$[A] \cdot \{T\} = \{f\}$$

with {T}, vector of internal forces and {f} vector of external actions on nodes. According to these values, spatial reticulate systems are categorised into four classes.

Table 5.1 *Classification of spatial reticulate structures*

Class	r_A	m, ss	Observations
1	$r_A = b$ $r_A = N$	ss = 0 m = 0	Statically and cinematically determinate systems
2	$r_A = b$ $r_A < N$	ss = 0 $m = N - r_A$	Cinematically indeterminate systems
3	$r_A < b$ $r_A = N$	$Ss = b - r_A$ m = 0	Statically indeterminate systems
4	$r_A < b$ $r_A < N$	$Ss = b - r_A$ $m = N - r_A$	Statically and cinematically indeterminate systems

Tensegrity systems are mainly class 4, and checking that they are stabilised by the state of self-stress requires the determination of the order of the infinitesimal mechanisms. These self-stress states have also to be compatible with the unilateral rigidity of cables.

For example, the case of the three-strut tensegrity system (twelve members, six nodes and twelve degrees of freedom) leads to:

E 5.51 $$ss = m = 1$$

The corresponding infinitesimal mechanism involves the translation and rotation of the upper triangle by respect to the bottom triangle (Figure 5.22).

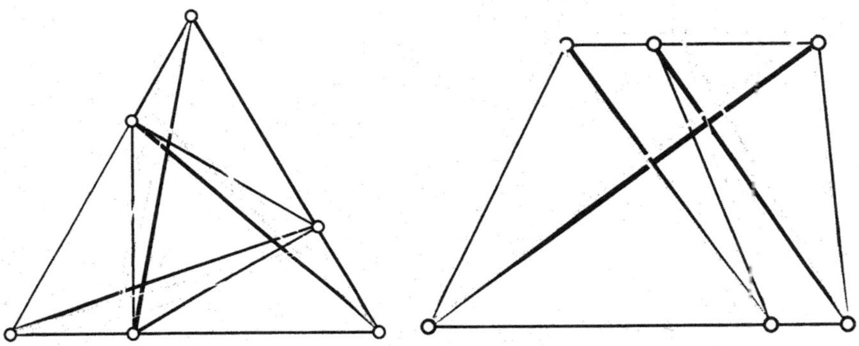

Figure 5.22 *Three strut tensegrity: infinitesimal mechanism*

5-4.3 Self-stress states

The stiffness of tensegrity systems results from states of self-stress. Mechanically speaking self-stress can be defined by the equilibrium equation:

E 5.52 $$[A]\cdot\{T\}=\{0\}$$

The right hand member of this equation is equal to a zero vector, since this matrix equation describes equilibrium of nodes without any external action.

Some simple examples of self-stressed systems are as follows:

- for the linear case, a cable put inside a tubular strut and fixed at their two common ends in such a way that the cable is tensioned and the strut compressed.
- for the planar case, the so-called "Saint Andrew" comprising two diagonal cables inserted in a quadrangular set of struts.
- for spatial cases, an octahedron whose edges are created with struts and the diagonals with cables (see Figure 3.22 in Chapter 3).

The specificity of these examples relies on the rigidity of tensioned elements, which cannot resist compression; it can be said that these members have "unilateral rigidity". An assembly containing such members is called a "tension system".

The spatial case mentioned illustrates the class of tension spatial reticulate self-stressed systems. Tensegrity systems constitute a subclass of these "tension systems", since they are spatial, reticulate, and self-stressed. Another main characteristic is the discontinuity of the set of struts and the continuity of the set of cables. From a theoretical point of view, the compressed elements of a tensegrity system only require unilateral rigidity, but this is technologically difficult to achieve and, of course, more expensive than using standard members, i.e. able to carry both tension and compression.

5-4.4 Mechanisms

5-44.1 Infinitesimal mechanism

Let us illustrate the mechanisms with simple examples that can be extended to tensegrity systems[4].

If we consider two nodes of a reticulate structure, namely "i" and "j", we call d_{ij} the geometrical distance between these nodes, and l_{ij} the manufacturing length of the element, which is inserted between these two nodes.

If a load can be applied to the system in such a way that the nodes move without modifying d_{ij}, for all couples of nodes, the corresponding displacement is called a *finite mechanism* (or "inextensional" mechanism). The system is cinematically indeterminate (i.e. its geometry is not defined). If the length variations are infinitesimal and of a lower order than the order of the displacement, the mechanism is called an *infinitesimal mechanism* and the system is also cinematically indeterminate.

Let us take two straight collinear assembled elements (Figure 5.23) verifying:

E 5.53 $$l_{12} = l_{23}$$

and

E 5.54 $$l_{12} + l_{23} = d_{13}$$

This leads to an infinitesimal mechanism (infinitesimal displacement which can be activated by an external action F); for a small displacement "d", the length variations are:

E 5.55 $$\Delta l = \alpha \cdot d^2 \leq d$$

[4] A first approach has already been presented in Chapter 3, "Fundamental concepts".

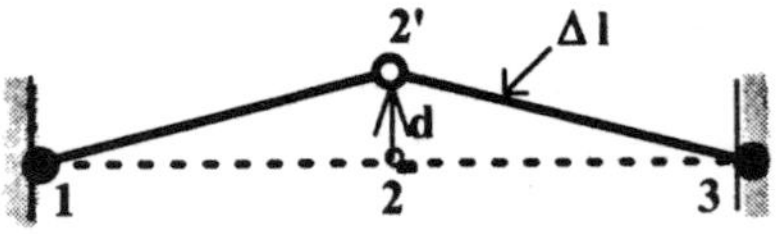

Figure 5.23 *Infinitesimal mechanism*

5-44.2 Stabilisation of infinitesimal mechanisms

Pre-stress or self-stress can stabilise certain infinitesimal mechanisms. In the previous example, two states of pre-stress can be introduced: tension or compression pre-stress.

Let us firstly introduce a tensile pre-stress (Figure 5.24) by choosing:

E 5.56 $$l_{12} = l_{23}$$

and

E 5.57 $$l_{12} + l_{23} < d_{13}$$

If we move node 2 from its initial location with an external action F, we activate the infinitesimal mechanism. When the action F is suppressed, node 2 comes back to its initial location, because during the displacement, the potential energy in the system has increased, and then decreases to the minimum level allowed by the boundary conditions.

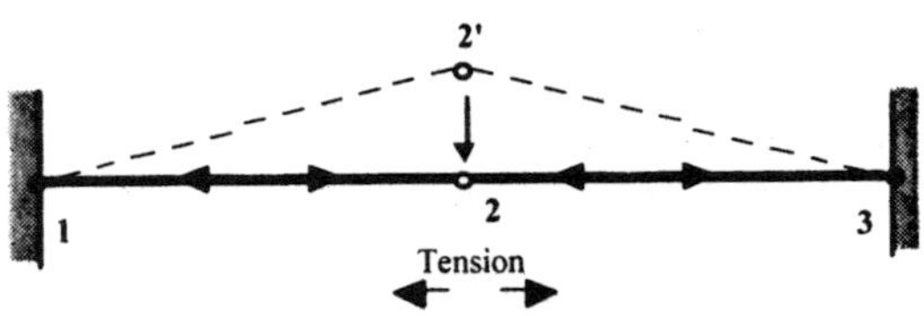

Figure 5.24 *Tension pre-stress*

We then introduce a compressive pre-stress (Figure 5.25) by choosing:

E 5.58 $$l_{12} = l_{23}$$

and

E 5.59 $$l_{12} + l_{23} > d_{13}$$

in order to build a system where members are compressed and aligned.

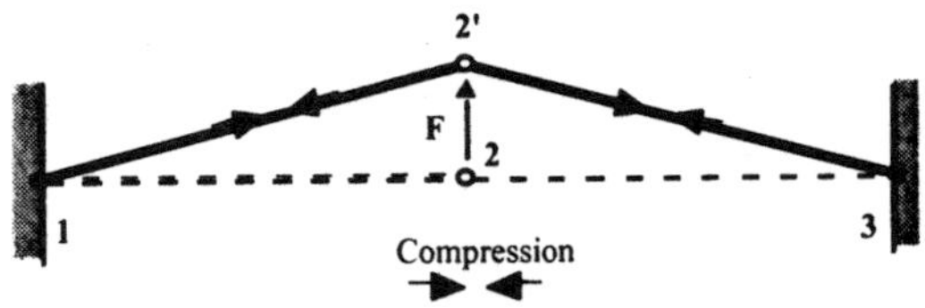

Figure 5.25 *Compression pre-stress*

When an external action F is applied, the infinitesimal mechanism is activated until a state where 2' becomes the new position of node 2, satisfying:

E 5.60 $$l_{12} = d_{12'}$$

and

E 5.61 $$l_{23} = d_{2'3}$$

Consequently, the members are unstressed, potential energy equals zero. When the action F is removed, the system stays in the displaced position; it does not return to its initial position, for which its internal energy would be higher.

The previous example illustrates the fact that, depending on the nature of the state of pre-stress applied, the infinitesimal mechanism is stabilised (with tension) or not (with compression). This is illustrated for plane pre-stress, but it is also true for self-stress states, and can be generalised to spatial structures, by the study of their equilibrium matrix and of the energy variations which are associated with the infinitesimal mechanism [Ref 5-1].

5-44.3 Assembly of cells

It is interesting to give an example of the evolution of "ss" and "m", when an assembly of cells is achieved.

5-443.1 Two cells

Let us begin with two triplex that are assembled in such a way that three nodes are common, consequently two cables are also common. We refer to it as "unilateral" assembly (Figures 5.26 and 5.27).

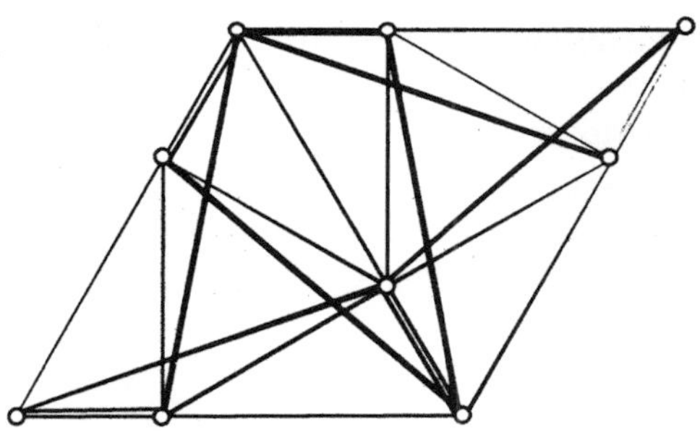

Figure 5.26 *Assembly of two triplex. Plane view*

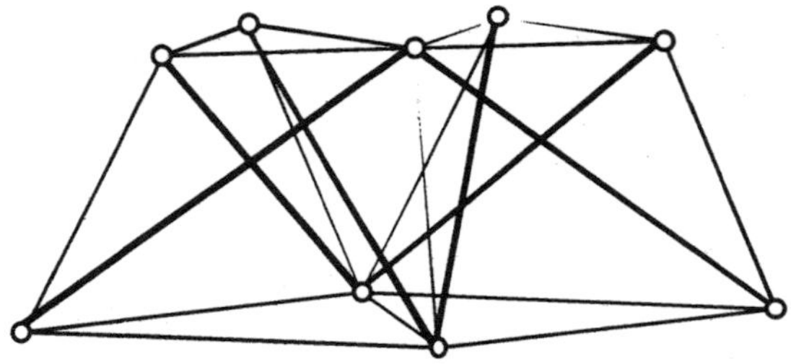

Figure 5.27 *Assembly of two triplex. Elevation view*

Taking account of relations that have previously given (we may write:

E 5.62
$$\left.\begin{aligned} b = 12 + 10 = 22 \\ n = 9 \Rightarrow N = 3n - 6 = 21 \end{aligned}\right\} \Rightarrow b - N = ss - m = 1$$

and consequently:

E 5.63
$$ss = m + 1$$

An inspection of the equilibrium matrix leads to two independent self-stress vectors and therefore one infinitesimal mechanism, which corresponds to the assembly of the two elementary infinitesimal mechanisms. These mechanisms are of order 1 and can be stabilised with a compatible[5] self-stress state.

When referring to remarks that have been written for a single cell, it is interesting to note that for an assembly with at least two cells, it is possible to exchange struts and cables without loosing stability. This means that the stability of one basic triplex (with three struts and nine cables) allows the permanence of stability for dual triplex. The word "dual" is used for two cells with an exchange of elements.

5-443.2 More than two triplex

Every unilateral assembly leads to only one mechanism and an increasing value of "ss", since with a supplementary cell the value of the difference between the number of elements "b" and the number of degrees of freedom "N" is increasing by one unit (Table 5.2).

[5] They respect the unilateral rigidity of cables.

Table 5.2 *Assembly of several regular triplex*

nb	Assembly	b	n	N	r_A	ss	m
1	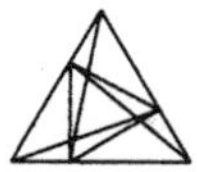	12	6	12	11	**1**	**1**
2	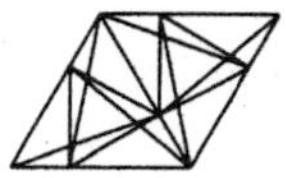	22	9	21	20	**2**	**1**
3	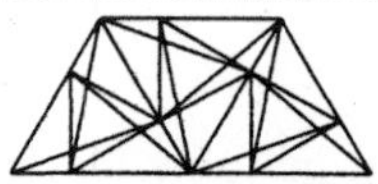	32	12	30	29	**3**	**1**
4	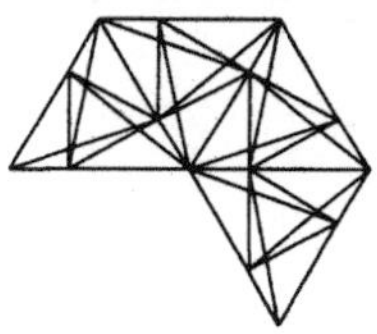	42	15	39	38	**4**	**1**
5	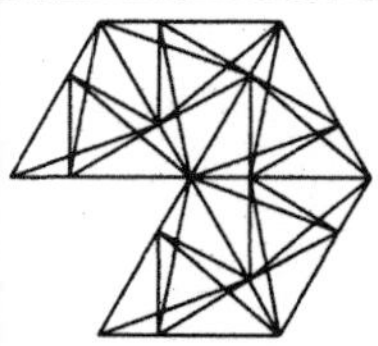	52	18	48	47	**5**	**1**
6	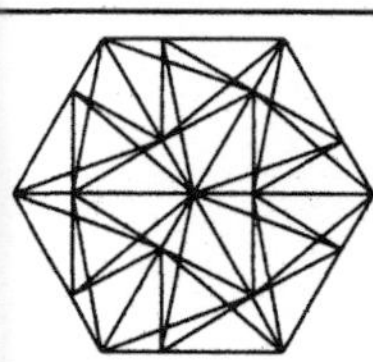	60	19	51	50	**10**	**1**
7	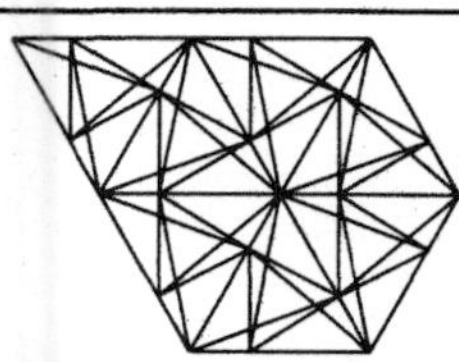	70	22	60	59	**11**	**1**
+1	Unilateral assembly	+10	+3	+9	+ ss = +m +1		
+1	Bilateral assembly	+8	+1	+3	+ ss = +m +5		

If we try to build an assembly of six cells, by including a sixth cell, we increase the number of elements by 8, and the number of degrees of freedom by 3; this is called a bilateral assembly. It can be demonstrated [Ref 5-1] that "ss" rises to 10 and "m" remains at value 1, keeping the relation:

E 5.64
$$\left.\begin{array}{r} +b = +8 \\ +n = +1 \Rightarrow +N = +3 \end{array}\right\} \Rightarrow +b-(+N) = +ss-(+m) = 5$$

Vassart has established that it is possible to eliminate infinitesimal mechanisms. It is necessary to destroy the regularity of the resulting hexagon in the case of six-cell assembly: therefore translation and rotation that are included in the remaining mechanism become impossible.

The evolution of self-stress states and mechanisms is given in Table 5.3: when the assembly is closed (six cells), "m" cancels.

Table 5.3 *Assembly of several irregular triplex*

nb	Assembly	b	n	N	r_A	ss	m
1	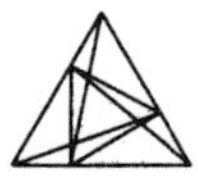	12	6	12	11	**1**	**1**
2	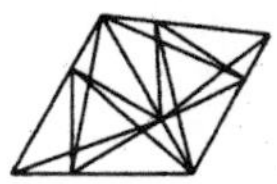	22	9	21	20	**2**	**1**
3	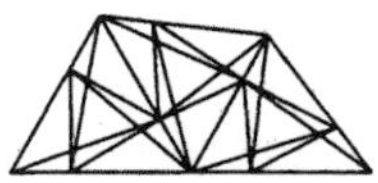	32	12	30	29	**3**	**1**
4	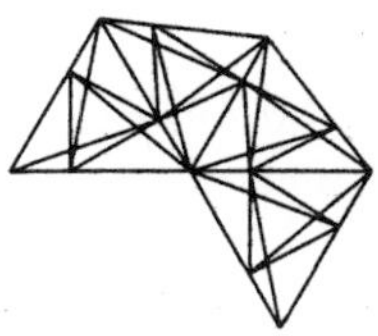	42	15	39	38	**4**	**1**
5	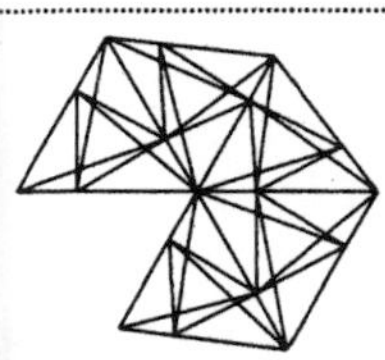	52	18	48	47	**5**	**1**
6	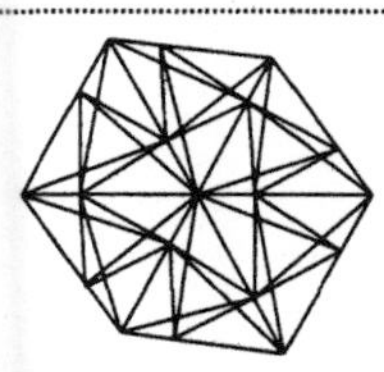	60	19	51	51	**9**	**0**
7	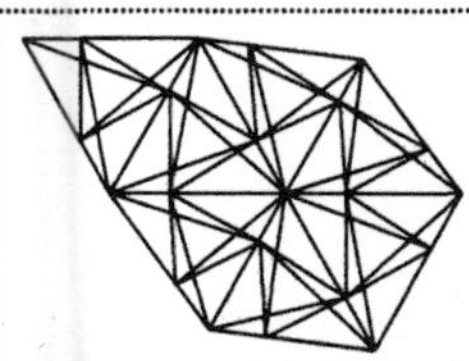	70	22	60	60	**10**	**0**
+1	Unilateral assembly	+10	+3	+9	+ss = +m +1		
+1	Bilateral assembly	+8	+1	+3	+ss = +m +5		

The example is illustrated in Figure 5.28. Six irregular triplex are assembled in such a way that the resulting configuration is not a regular hexagon assembly.

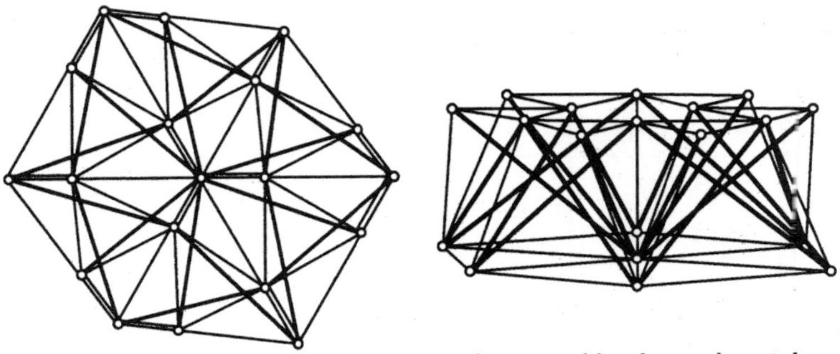

Figure 5.28 *Plane and perspective view of an assembly of irregular triplex*

5-4.5 Conclusion

Based upon our own experience, it appears that the control of self-stress is a main feature of tensegrity systems and specific attention is required in order to identify self-stress states, to choose an adequate combination and to succeed in their effective implementation. Infinitesimal mechanisms, if any, can be stabilised by these states of self-stress, and also by specific boundary conditions in real structures.

5-5 Self-stress qualification

5-5.1 Self-stress determination

In previous sections we described the force density method, which can be used in the form-finding process. Keeping the same notations and basic definitions, the force density is:

E 5.65 $$q_j = \frac{T_j}{l_j^0}$$

We have defined a self-stress state as a vector of force density noted $\{q^0\}$, that verifies the equilibrium equation:

E 5.66 $$[A] \cdot \{q^0\} = \{0\}$$

with [A], being the equilibrium matrix. This equation with a zero right hand-side vector defines the self-stress condition (internal stresses without any external actions). $\{q^0\}$ is a b-line vector. The number of self-stress states "s" is the dimension of the vectorial kernel subspace of the equilibrium matrix and noted "ker A". The value of "ss" was soon already given in previous sections (see E 5.47).

Then we can write that:

E 5.67 $$\{q^0\} \in \ker(A)$$

The numerical determination of the "ss" vectors is a classic mathematical problem, which can be solved for instance by using augmented matrices or generalised inverse methods (see [Ref 5-11]).

ss independent vectors define a self-stress basis [S]:

E 5.68 $$\{q^{0,k}\} \rightarrow k = 1,...,ss$$

In order to simplify matters, we will omit the superscript "0", since no confusion occurs.

Consequently a basis [S] is such as:

E 5.69 $$[S] = \begin{bmatrix} q_1^1 & q_1^2 & \dots & q_1^{ss} \\ \dots & \dots & \dots & \dots \\ \dots & \dots & \dots & \dots \\ q_j^1 & q_j^2 & \dots & q_j^{ss} \\ \dots & \dots & \dots & \dots \\ \dots & \dots & \dots & \dots \\ q_b^1 & q_b^2 & \dots & q_b^{ss} \end{bmatrix}$$

It is useful for subsequent mathematical calculations to order the basis so as the first "t" lines correspond to the tensioned components and the "b-t" other lines to the compressed components. Any linear combination of the basis vectors is a self-stress state:

E 5.70 $$\{q_{si}\} = \alpha_{si}^1 \cdot \{q^1\} + ... + \alpha_{si}^k \cdot \{q^k\} + ... + \alpha_{si}^{ss} \cdot \{q^{ss}\}$$

with k =1,...,ss

5-5.2 Feasible self-stress state

According to their definition there can be an infinite number of self-stress states. But the associated vectors must satisfy a necessary condition, which is related to the unilateral rigidity of tensioned components. This is a feasibility condition and in turns leads to the "feasible" self-stress state. This is not ensured by mathematical treatment, which does not take into account the sign of the vector components. An appropriate choice of coefficients α_{si}^k ("si" being used for the self-stress state $\{q_{si}\}$) is then required. Taking into account the numbering order previously described and the sign convention of positiveness for tension, a set of b conditions can be written.

➢ "t" conditions for tensioned components:

E 5.71
$$\begin{aligned}&\alpha^1_{si}\cdot q^1_1+\alpha^2_{si}\cdot q^2_1+\ldots+\alpha^k_{si}\cdot q^k_1+\ldots+\alpha^{ss}_{si}\cdot q^{ss}_1\geq 0\\&\ldots\ldots\ldots\ldots\ldots\ldots\ldots\ldots\ldots\ldots\\&\alpha^1_{si}\cdot q^1_t+\alpha^2_{si}\cdot q^2_t+\ldots+\alpha^k_{si}\cdot q^k_t+\ldots+\alpha^{ss}_{si}\cdot q^{ss}_t\geq 0\end{aligned}$$

➢ And "b-t" conditions for compressed components:

E 5.72
$$\begin{aligned}&\alpha^1_{si}\cdot q^1_{t+1}+\alpha^2_{si}\cdot q^2_{t+1}+\ldots+\alpha^k_{si}\cdot q^k_{t+1}+\ldots+\alpha^{ss}_{si}\cdot q^{ss}_{t+1}\leq 0\\&\ldots\ldots\ldots\ldots\ldots\ldots\ldots\ldots\ldots\ldots\\&\alpha^1_{si}\cdot q^1_b+\alpha^2_{si}\cdot q^2_b+\ldots+\alpha^k_{si}\cdot q^k_b+\ldots+\alpha^{ss}_{si}\cdot q^{ss}_b\leq 0\end{aligned}$$

The problem to solve is to find a set of α^k_{ss} satisfying these conditions. This is a known problem of *linear programming*, when the whole set is transformed so that all the conditions are positiveness conditions. The method of solution is the "*simplexe*" method [Ref 5-12].

As an example let us consider an assembly of three four-strut cells (Figure 5.29):

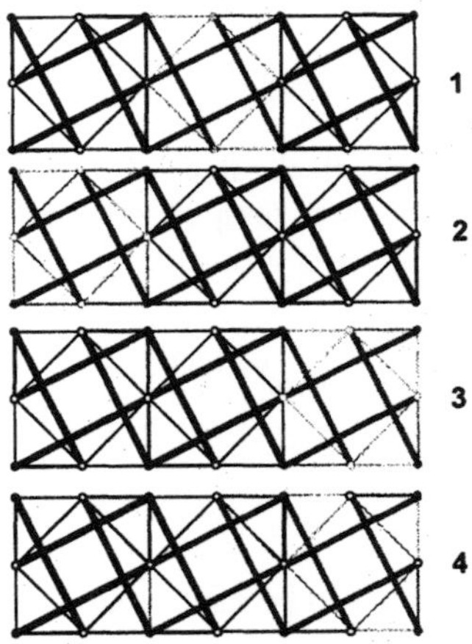

Figure 5.29 *Feasible independent self-stress states*

The study of the corresponding equilibrium matrix leads to four self-stress basis vectors. It is then necessary to find four independent self-stress states. Three of them are partial states that are derived from the structure itself, since they fit exactly with the constitutive cells. The fourth one is not so obvious: one possibility is shown on step 4 of Figure 5.29. Two cells are concerned in this case.

5-5.3 Existence of feasible self-stress

Then feasibility of self-stress states is defined by the set of conditions related to tensioned components (E 5.71). The problem of existence can be illustrated in the case of the low values of ss.

When ss = 1, the verification is reduced to the direct sign inspection for the force density coefficients of tensioned components since E 5.71 becomes:

E 5.73
$$\alpha_{si}^{1} \cdot q_{1}^{1} \geq 0$$
$$\ldots\ldots\ldots\ldots$$
$$\alpha_{si}^{1} \cdot q_{t}^{1} \geq 0$$

A second inspection can be done for ss = 2. In this case we have to check a set of inequalities of the following form:

E 5.74
$$\alpha_{si}^{1} \cdot q_{1}^{1} + \alpha_{si}^{2} \cdot q_{1}^{2} \geq 0$$
$$\ldots\ldots\ldots\ldots\ldots\ldots\ldots\ldots$$
$$\alpha_{si}^{1} \cdot q_{t}^{1} + \alpha_{si}^{2} \cdot q_{t}^{2} \geq 0$$

Each row of this set of conditions is associated with one of the "t" tensioned components. The nullity can be plotted as a straight line in a plane, which defined by the variables $\alpha_{si}^{1}, \alpha_{si}^{2}$. It is then possible to check the existence of feasible self-stress. Three possibilities occur:

- Several values of $\alpha_{si}^{1}, \alpha_{si}^{2}$ can be chosen inside a sector defined by two straight lines D_i and D_j (in white in Figure 5.30).

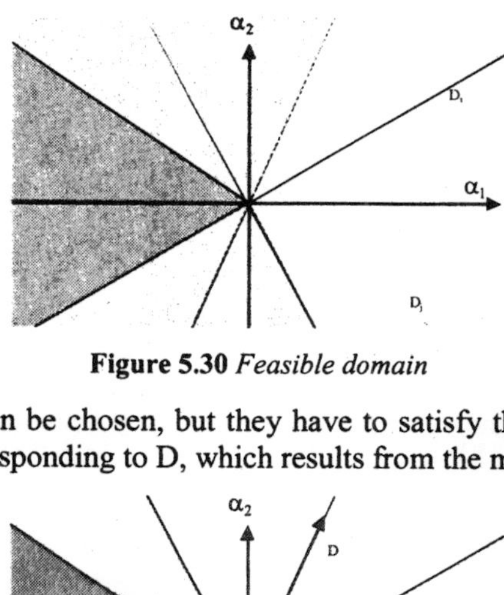

Figure 5.30 *Feasible domain*

- Several values can be chosen, but they have to satisfy the equation of a single straight line corresponding to D, which results from the merging of D_i and D_j.

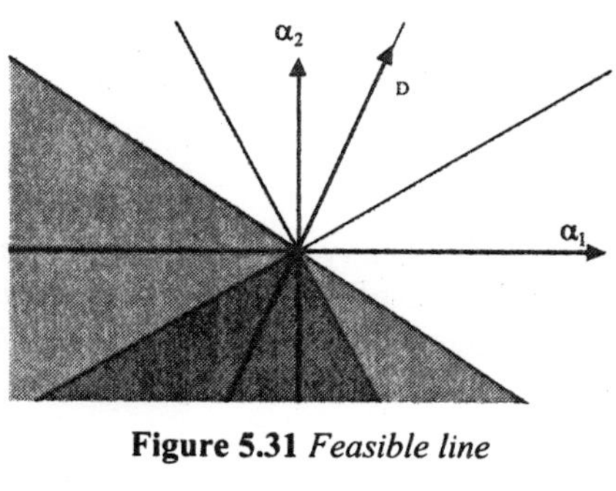

Figure 5.31 *Feasible line*

There is no feasible self-stress when no acceptable domain or line remains, after applying the set of inequalities E 5.74 (see Figure 5.32).

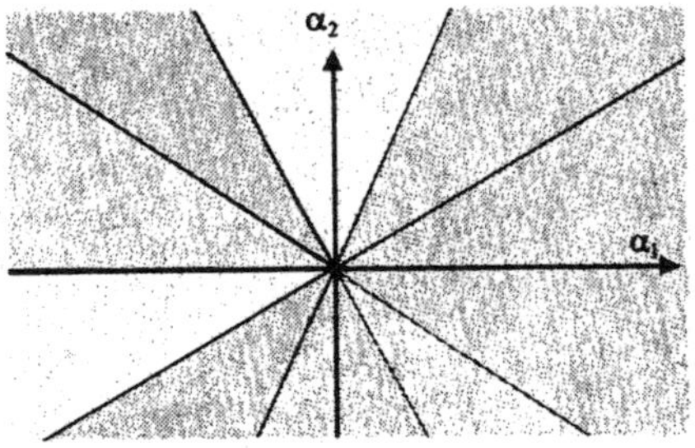

Figure 5.32 *No feasible self-stress*

These remarks can be generalised [Ref 5-12]. In the case of ss = 3, a feasibility domain can be defined.

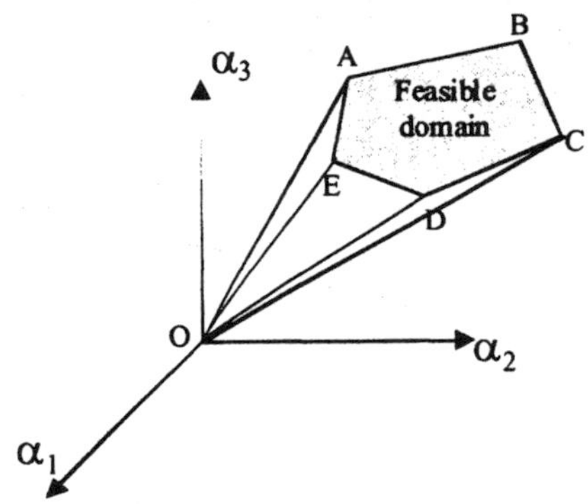

Figure 5.33 *Feasibility domain for ss = 3*

5-5.4 Form and forces

Tensegrity systems belong to those specific systems for which the coupling between geometry and stresses is a main parameter. When dealing with self-stress problems it is sometimes interesting to geometrically qualify a self-stress state. When self-stress states are partial, they also can be geometrically localised like those described in Figure 5.29. This can be called a "local self-stress state" (which is consequently partial). In the case of the above example we described four independent local self-stress states, three of them corresponding to elementary four strut cells, which have been aggregated to constitute this system.

5-5.5 Conclusion

Self-stress is a key feature of tensegrity systems. It must be studied with special care not only to make an optimum choice of the initial state, but also in accordance with practical aspects for implementation and monitoring.

5-6 Designing tensegrity systems

5-6.1 Introduction

The design of tensegrity systems basically goes through three stages. Starting from a given configuration, the self-stress states likely to be used as soon as the system is assembled must firstly be determined. Then, the choice of self-stress state having been made, the components must be designed. And, if necessary, the study of structure sensitivity to inaccuracies in the fabrication of the elements should allow us to be certain of its stability. Since self-stress problems have been largely developed in the previous sections, we now turn to the description of some features relating to design.

5-6.2 Mechanical behaviour of tensegrity systems

5-62.1 Geometrically non linear behaviour

We developed a first study by taking into account a numerical model with geometric non-linearity. This model is based on an incremental writing of the principle of virtual work according to a total Lagrangian formulation. Even if a hypothesis of linear elasticity is used in what follows, it is possible to develop the same model with material non-linearity. This leads to a system of non linear equations, where the tangent stiffness matrix $\left[{}^{t}_{0}K_{b}\right]$ is linked with internal stress vector $\left\{{}^{t}_{0}F\right\}$:

E 5.75 $$\left[{}^{t}_{0}K_{b}\right]\cdot\left\{{}^{\tau}_{t}u\right\}=\left\{{}^{\tau}R\right\}-\left\{{}^{t}_{0}F\right\}$$

with

E 5.76 $$\left[{}^{t}_{0}K_{b}\right]=\left[{}^{t}_{0}K_{LA}\right]+\left[{}^{t}_{0}K_{LB}\right]+\left[{}^{t}_{0}K_{NL}\right]$$

In this expression $\left[{}^{t}_{0}K_{LA}\right]$ is the classic linear rigidity matrix, which is used for the case of small displacements; $\left[{}^{t}_{0}K_{LB}\right]$ is the initial displacement matrix which takes into account the non linearity, and is related to the quadratic terms of complete strain expressions. $\left[{}^{t}_{0}K_{NL}\right]$ is the geometric matrix, or initial stresses matrix, resulting from the non-linear strains. $\left\{{}^{\tau}R\right\}$ is the vector of external actions and $\left\{{}^{\tau}_{t}u\right\}$ the incremental displacement vector between two instants t and τ.

The expression E 5.75 is representative of a non-linear system of equations and the solving method is necessary incremental. This iterative method requires for each step a correction of variables in order to ensure convergence.

The basic scheme of study is illustrated in Figure 5.34. It corresponds to an incremental model.

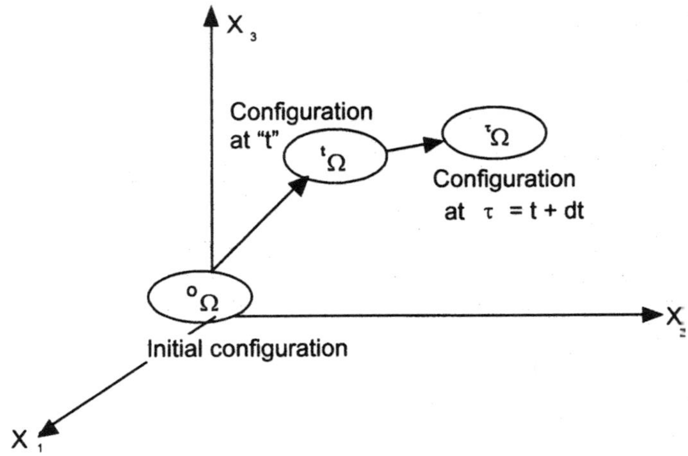

Figure 5.34 *Non-linear process*

This process is used, in the first instance, to simulate the introduction of self-stress, by imposing the external action vector to be identically equal to zero, and searching for a target state of self-stress. It is then possible to load the system with external actions and to take the final stage of the search for self-stress as initial conditions for the second application of the iterative process.

5-62.2 Mechanical behaviour of the four strut module ("quadruplex")

As an example let us provide some information about the mechanical behaviour of a *quadruplex,* which is a simple four-strut module. Six conditions are imposed on nodes 1, 2 and 3. Node 1 is completely fixed, node 2 can only move along OY', and node 3 can move in plane Y'OZ' (see Figure 5.35). External loads are applied on nodes 6, 7 and 8. According to the direction of these actions several cases are defined: compression, traction, upward flexion, downward flexion and torsion. The results of calculations are plotted in Figures 5.36 to 5.38.

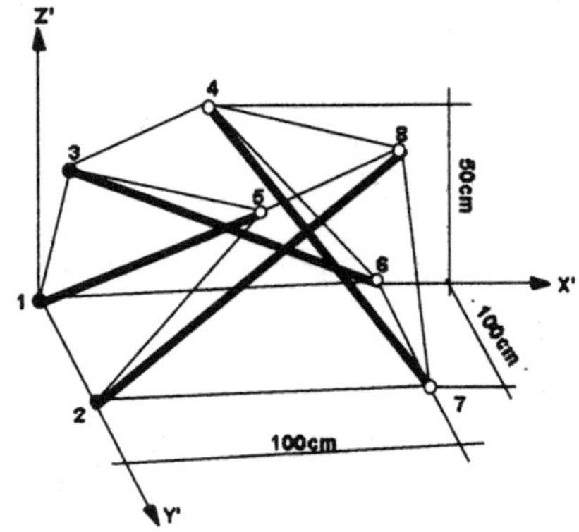

Figure 5.35 *Quadruplex*

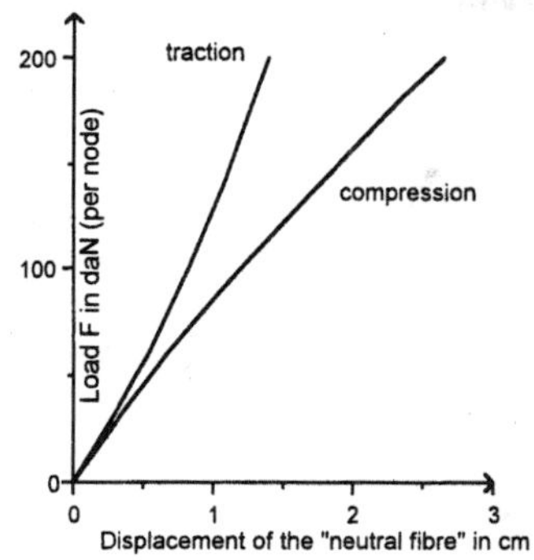

Figure 5.36 *Compression and traction*

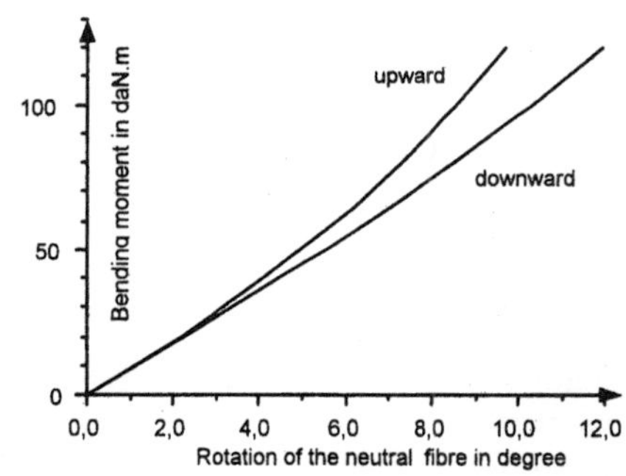

Figure 5.37 *Flexion*

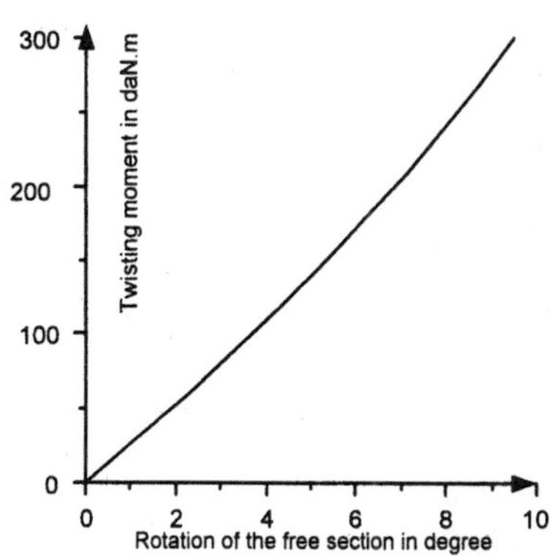

Figure 5.38 *Torsion*

Some basic characteristics of mechanical behaviour are illustrated in these graphs: non linearity for each case, anisotropy between traction and compression, the same in flexion according to the sense (upward and downward). It is also obvious that in many cases external loading rigidifies the system (resulting from a non linear analysis). This is a very simple example; tensegrity grids with more than one hundred cells have been tested with our specific software ("*Tensegrité 2000*").

5-6.3 The design

Considering the numerous parameters that come into play in the behaviour of tensegrity systems, their design is difficult. Moreover, as their behaviour is not geometrically linear, a second order study is necessary [Ref 5-19]: the half cuboctahedron [Ref 5-20] or the expanded octahedron [Ref 5-21]. These studies have highlighted the different parameters linked to the design: level of self-stress, distribution of the self-stress (geometrical aspect), and rigidity ratios between compressed and tensioned components.

These parameters will have an action both on the overall stiffness of the structure, and at the same time on the stress obtained under load. Consequently, the design of tensegrity systems is an iterative process.

5-63.1 Design criteria

As the design parameters are numerous, it is important to give simple and accurate criteria allowing for the fastest design of a structure.

In the case of grids for roofing, for example, the problem is simplified: since the maximum displacements are deliberately limited according to the span (generally at 1/200th), the behaviour is not far away from a first order calculation (service state). The level of self-stress has thus little effect on the stiffness of the grid: action needs to be taken on the cross-sections of elements to meet the deflection criterion. The second-order calculation, however, still remains necessary for ultimate states. Moreover, it must be emphasised that in the case of equal self-stress, the tension in the elements varies only slightly (as opposed to deflection) in changing the rigidity ratio between the struts and the cables: this result is similar to that obtained in isostatic structures. This will be of help in the choice of the self-stress level of the structure, which is always defined within one multiplication factor.

5-63.2 Design calculations

The design process we propose is thus generally in two stages:

- A service state design ensures that the deflection criterion is met, while remaining within the acceptable limits for the stress in the elements. Moreover, as tensegrity systems are tension structures, we ensure that none of the cables present in the structure be slack.
- An ultimate design state verification ensures the overall stability of the structure under extreme loading. Self-stress is a permanent action with both acting and resistance characteristics at the same time. Thus, when the ultimate design state is carried out, both aspects must be taken into account. The former will reduce self-stress to check that the structure keeps an overall stability ("resistant" self-stress), the latter increases self-stress to check the local stability of elements

("acting" self-stress). Determination of partial safety coefficients is detailed in [Ref 5-21].

A rigidity ratio between struts and cables and a self-stress level must first of all be specified before starting on a first calculation. For sufficient rigidity, our experience in this field has shown that a rigidity ratio (EA_{struts}/EA_{cables}) close to 10 is satisfactory. Above this, the behaviour is too flexible and leads to over sizing the cable elements. Below 10, the struts are overloaded and thus oversized.

5-6.4 Sensitivity problems

The problem of sensitivity can be illustrated using a very simple example, an isolated cable. We will see that the consequences of a very slight variation in the length of the element will be important in terms of resulting stress.

Imagine that a cable is to be linked to two fixed points with a given stress target. Using Hooke's law, it is possible to recalculate the manufacturing length to obtain the expected tension *in situ*. Note that so far we are not interested in the relaxation aspect. Let us take a practical example (Figure 5.39). If we assume that Young's modulus "E" is equal to 125 000 MPa, with a cross section area "A" of 0,5 cm^2, we may write:

E 5.77 $$T = E.A.\frac{\Delta l}{l^{free}} = E.A.\frac{l^{0} - l^{free}}{l^{free}}$$

With l^{free}, being the manufacturing cable length. For pre-stress target of 20 kN, the required value of l^{free} is equal to 0,9968 m.

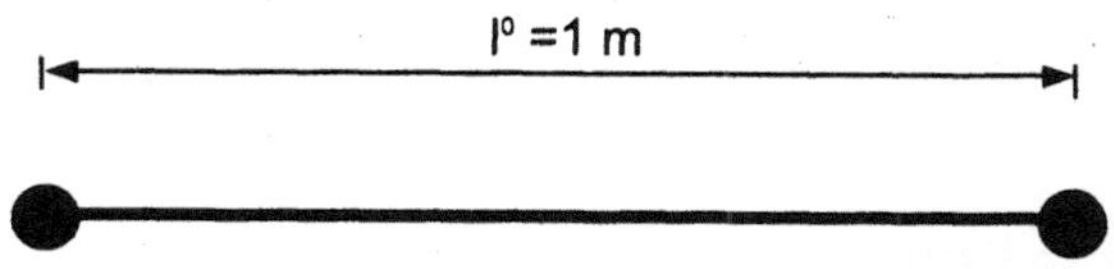

Figure 5.39 *Pre-stressed cable between two fixed points*

This requires a lengthening of about 3 mm. If the accuracy of manufacture of the cable length is about ±1mm, the consequences on the tension *in situ* will be important.

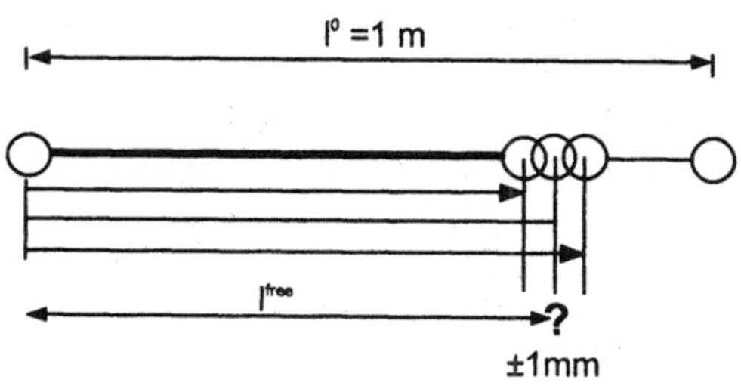

Figure 5.40 *Actual manufacturing length of the cable*

In the case of this example, a manufacturing tolerance of the cable of ±1 mm induces a variation in the tension of ±6kN. This is considerable and unacceptable from a constructive point of view, and we have to quantify the influence of these manufacturing errors on the whole of the structure. But since the behaviour of tensegrity systems is not linear and the random variables are numerous, analytical deterministic treatment is impossible. To carry out this study, a certain number of Monte-Carlo simulations must be made, and then the result treated statistically.

The disturbances in length, which are introduced in each of the elements come from the realisation (probabilistic concept described in [Ref 5-22]) of a random variable, the statistical parameters of which will depend on the manufacturing accuracy expected for the elements. Considering the manufacturing processes, we will take into account a certain manufacturing accuracy for the cables, the struts being supposed to have the required length.

It is then possible, for each simulation, to recalculate the equilibrium using this new data. Using the results obtained, a statistical treatment will define the characteristic minimum and maximum values between which the tensions can almost certainly be found [Ref 5-22]. These characteristic values can be defined with some certainty considering the discrete nature of the sample.

Then, all we need to do is to check that the characteristic values are located in the interval defined by two cases of ultimate loading state to justify the partial safety factors. These partial safety factors have been chosen to be equal to 0.8 and 1.2 depending on whether self-stress can be considered as favourable for equilibrium or not [Ref 5-12] (no precise rules exist in this domain and we chose these values for a first estimation, knowing that no cable slackening occurs during the corresponding loadings). If it is not so, the manufacturing tolerance of self-stress can be called into play.

5-6.5 Application to a tensegrity grid

To fix a realistic study framework, we studied a square grid 9 meters long constituted of 36 half-cuboctahedron modules which has already been described. This is a classical example of aggregation of self-equilibrated modules. The height was determined so as to have the bars inclined at 45° with respect to layers. The height of the grid is then 1m15 giving an aspect ratio of about 1/8th. The grid thus assembled has 133 nodes and 516 components (144 bars and 372 cables).

5-65.1 Finding the self-stress states

The mechanical study of this statically and cinematically indeterminate structure shows 144 self-stress states, i.e. 144 different internal possibilities for submitting the grid to internal forces as soon as it is assembled. Of course, these states must be sorted so that the chosen self-stress be easy to apply on assembly. They also have to be feasible (compatibility with unilateral rigidity).

In the case of the half cuboctahedron, the problem is much simplified since once the modules have been juxtaposed, they maintain a local self-stress state, so a state of overall self-stress can be obtained simply by summoning the states particular to each module (Figure 5.41).

The choice of the self-stress is then made according to both its distribution and its level. We will choose a homogenous self-stress for easier assembly. As for its level, we will put ourselves at 50% from the limit recommended by European Code 3 (steel constructions) for the compression of struts.

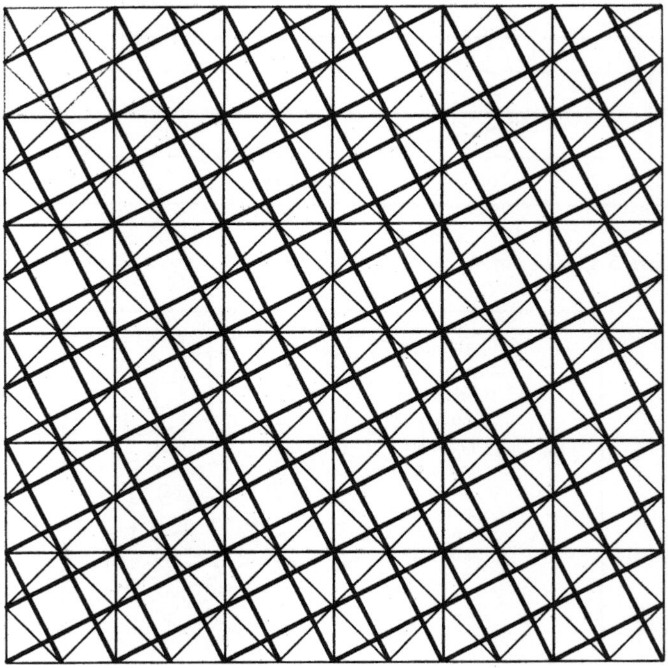

Figure 5.41 *State of self-stress localised on a single module (top left-hand square)*

5-65.2 Service Limit State design (SLS)

Altogether, the external load to be taken into account for the SLS is 135 daN/m^2, which includes a self weight of 25 daN/m^2 and a live load of 110 daN/m^2. The loads are supposed to be downwards and applied only to the nodes of the upper layer of the grid.

Before calculation, the mechanical characteristics of the elements must be fixed: a Young's modulus of 200 000 MPa for the struts and 125 000 MPa for the cables are ordinary values for steel elements.

With these conditions, we get a cross-section design of 4.14 cm^2 for the struts and of 0.654 cm^2 for the cables. Lower strut cross sections gave us displacements, which are incompatible with the deflection criterion (L/200). These characteristics correspond to the manufactured products: struts of 48.3 mm external diameter and 2.9 mm thickness and a multi-strand cable of 7x19 wires of 10 mm nominal diameter.

The following figures present the results in terms of deflection. Figure 5.42 gives the maximum deflection for the centre row of nodes (the most deflected) whereas Figure 5.43 presents the general rate of displacement of the nodes of the lower layer.

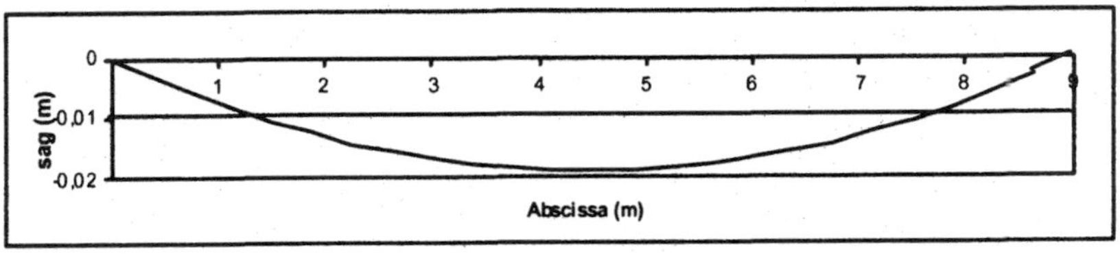

Figure 5.42 *Value of the vertical displacements for the centre row of nodes*

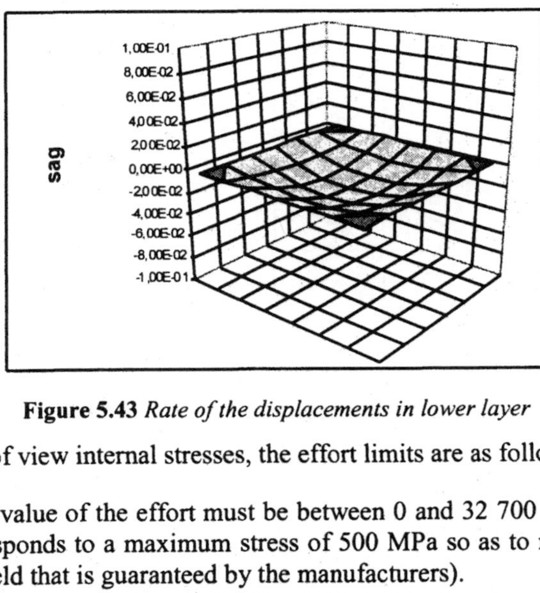

Figure 5.43 *Rate of the displacements in lower layer*

From the point of view internal stresses, the effort limits are as follows:

- Cables: the value of the effort must be between 0 and 32 700 N (the maximum value corresponds to a maximum stress of 500 MPa so as to remain within the elasticity field that is guaranteed by the manufacturers).
- Struts: the maximum value admitted by EC3 is linked to the value of the inertia of the strut, its length and its cross section. In our case, the cross section is of 4.14 cm^2 and tubular, the length of the struts is 2m03, inertia 10.7 cm^4. The maximum authorised value of internal force for 235 grade steel is 39 400 N.

Figure 5.44 shows the value of the tensions in the elements before and after loading. The chosen self-stress is a uniform self-stress, for which all the struts are loaded to

50% of their limit according to EC3, i.e. 20 000 N. The struts are numbered from 1 to 144 and the cables from 145 to 516.

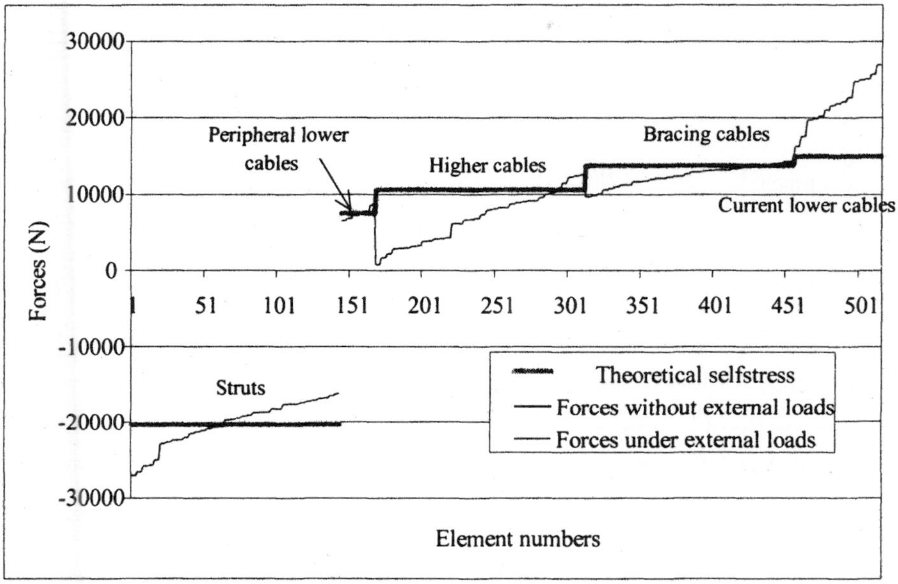

Figure 5.44 *Value of the stress in the elements before and after loading*

The criteria of stress limit are met, even if some cables are very close to a zero value. Quite logically, we note that the centre cables of the upper layer are the least tensioned. An increase in the level of self-stress could be considered, but increasing the level of stress in the elements can make them exceed their limits of stability criteria. So it is preferable to approach the limit by slackening a cable rather than exceeding the elasticity limit which would modify the behaviour of the structure in the long term.

5-65.3 Ultimate Limit State verificatio n

Now that SLS (Service Limit State) design has been completed, ULS (*Ultimate Limit State)* verification remains to be carried out. For this, we will apply the two combinations of actions (partial safety factors have been chosen to be equal to 0.8 and 1.2 depending on whether self-stress can be considered as favourable for equilibrium or not [Ref 5-12], (Figures 5.45 and 5.46).

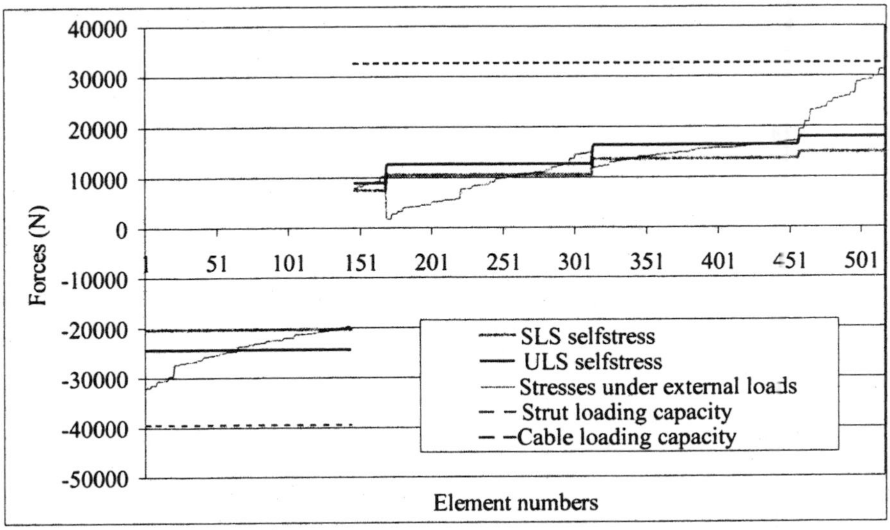

Figure 5.45 *ULS verification for self-stress equal to 1.2 times the expected self-stress*

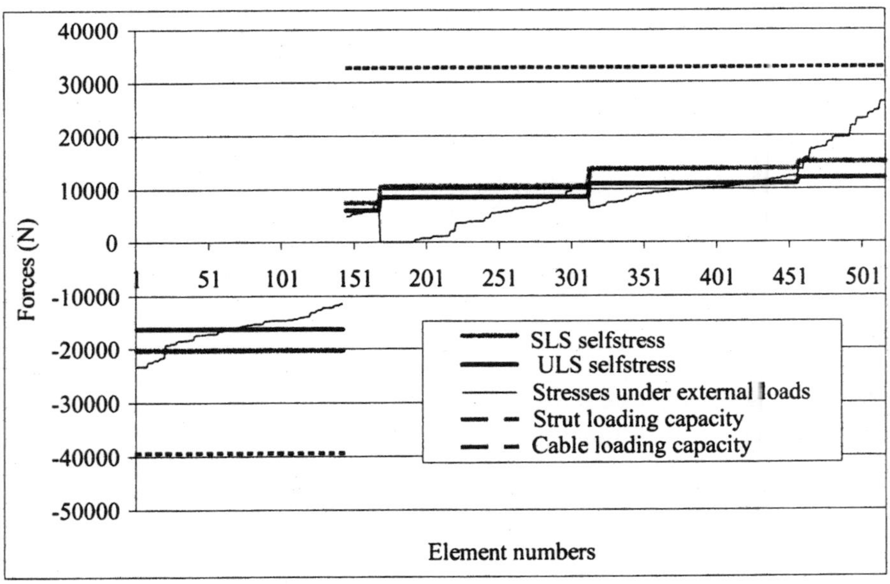

Figure 5.46 *ULS verification for self-stress equal to 0.8 times the anticipated self-stress*

As might be expected, the case of loads increasing the self-stress brought us to reach the stress limits of the elements, whereas the case of loads decreasing the self-stress slackened some cables.

If the first case does not bring the local and overall stability of the structure into question, it must be emphasised that in the second case the slackening of some cables did not ruin the structure. It maintains overall stability.

It is interesting to see the distribution of the internal forces at the end of loading. We thus note that their distribution is not homogenous at the end of loading (Figure 5.46). We can then see that the centre struts are (on average) less stressed than the surrounding ones and that the slack cables are to be found in the centre of the grid. This is the reason for suggesting adjusting the self-stress in order to make the best use of the elements.

The graphs show that moderate variations in the self-stress can significantly modify the behaviour of the structure under load. This variability in the manufacturing lengths appears to us to be the dominant factor in self-stress uncertainty. We shall thus present a study of the sensitivity of the grid to manufacturing inaccuracies in order to verify the relevance of the partial safety factors used for the ULS self-stress.

5-65.4 Study of the sensitivity of the grid

We carried out a statistical analysis of the tensions in the structure, for a simulation of fifty grids, with manufacturing accuracy of the cables disturbed by the realisation of a centred Gaussian distribution and a standard deviation of 0.51 mm. In this case, the disturbances constitute a sample of which 95% of the population is in the range of ± 1 mm.

The first study is concerned with self-stress. It enables the definition, for each of the elements, of an interval in which the stress will probably be found after the assembly of the structure.

After statistical treatment, an average value of self-stress (very) slightly different to that expected is obtained. Figure 5.47 presents the minimum and maximum characteristic values at 5% with 95% confidence.

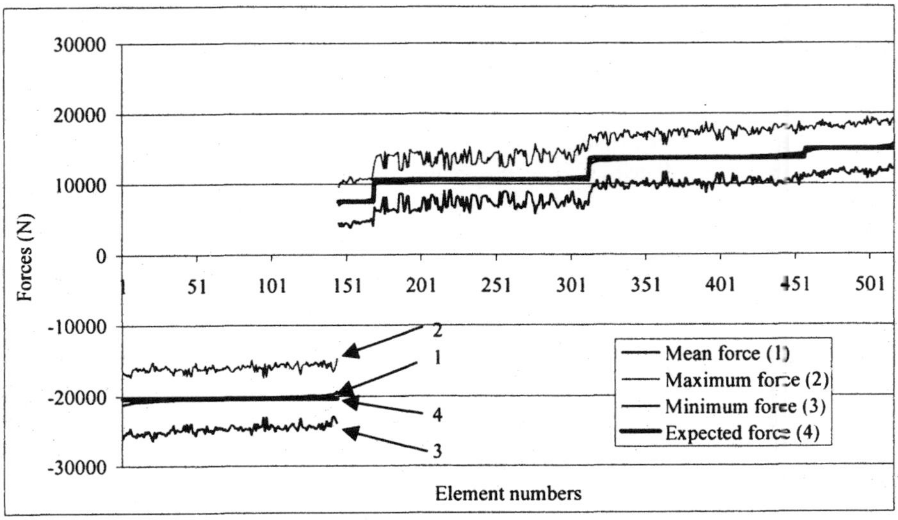

Figure 5.47 *Characteristic values at 5% stress with 95% confidence. Uniform state of self-stress. Statistical analysis on 50 grids*

In order to verify the relevance of the partial safety factors recommended for ULS self-stress, we have to analyse the variations in the ratio between the minimum and maximum characteristic values and the expected value of the stress. For the struts, the extreme values of this ratio equal 0.73 and 1.29 and for the cables 0.58 and 1.45, values which are far from the recommended safety factors (0.8 and 1.2). However, if we do not look at the extreme values but at the average ones, these factors are then estimated at 0.79 and 1.2 for the struts and 0.73 and 1.27 for the cables, values that can be considered close to the recommended ones.

These results confirm that the partial safety factors equal 0.8 when the self-stress is resistant, and 1.2 if it is acting load, constitute the first realistic approach if we consider that the manufacturing accuracy nowadays is of the order of one millimetre. We have also to keep in mind the other hypothesis (namely the initial self-stress level). However, better adequacy could be obtained either by extending the partial safety factors slightly or by imposing still better manufacturing accuracy.

5-65.5 Conclusion

After studying the different parameters that play a role in the behaviour of tensegrity systems, we took an interest in their design according to the Eurocodes and more particularly in the partial safety factors which affect the self-stress for verifications of ultimate limit state stability. A study of the stress element sensitivity to manufacturing and assembly accuracy enabled us to back up the proposed values of partial safety factors.

5-7 Active control

5-7.1 Introduction

Lightweight structures and more particularly those of the truss-cables type is of interest in numerous fields of application, such as civil engineering, mechanics - but also in aerospace science too.

Under external excitations, these structures are likely to present important vibrations. This observation leads to non-linear geometrical behaviour, and consequently, entails a modification of the structure stiffness in accordance with the displacements of its nodes.

Some scientists have studied many active control strategies. Active control was first proposed by Yao [Ref 5-23], and was later developed by numerous authors such as Abdel-Rohman ([Ref 5-24] [Ref 5-25] and [Ref 5-26]). Active control is an efficient means, which is currently creating an increasing amount of interest.

Many algorithms have been studied in the linear domain. We may quote the instantaneous optimal control algorithm presented by J. N. Yang ([Ref 5-27] [Ref 5.28] [Ref 5-29] and [Ref 5-30]), L. Chung [Ref 5-31] and F. R. Rofooei [Ref 5-32]. In these studies, the algorithm is based on a space-state representation and is well adapted to random excitation signals. By introducing Newmark's method, C. C. Chang [Ref 5-32] managed to express this algorithm in a non-space state representation. This particular writing enabled him to deal with the problems by using a finite element calculation, which has the advantage of ensuring better accuracy of results.

Moreover, other studies have dealt with the formulation of algorithms adapted to non-linear structures. Reinhorm and al. [Ref 5-34] have developed an active control algorithm for flexible structures. At the same time, T. T. Soong [Ref 5-35] and J. Z. Cha [Ref 5-36] have proposed numerical solutions based on the "descent" method (like the conjugate gradient method).

The present approach developed by S. Djouadi [Ref 5-37], from our own laboratory, proposes a method formulated from the instantaneous optimal control algorithm, which is applied to structures presenting effects of large displacement. A finite element approach is put forward with a hypothesis of material linearity in order to asses stiffness variation during displacements. This approach has been recently published [Ref 5-38].

The simulation of the behaviour of a three-dimensional beam representing the mast of an antenna validates the method. It is built with several assembled basic units, which are themselves considered as self-stable structures composed of compressed struts and tensioned cables.

5-7.2 Instantaneous optimal control

If we assume that the mass M and the damping C are constant, then the modelling of the dynamic behaviour of a controlled structure under non-linear geometrical conditions can be formulated by the following differential equation:

E 5.78 $$M\cdot {}^{\tau}\!\left(\frac{\partial^2 x}{\partial t^2}\right) + C\cdot {}^{\tau}\!\left(\frac{\partial x}{\partial t}\right) + {}^{\tau}F = {}^{\tau}F_{ext} + {}^{\tau}B\cdot {}^{\tau}F_c$$

Vectors ${}^{\tau}\!\left(\frac{\partial^2 x}{\partial t^2}\right)$ and ${}^{\tau}\!\left(\frac{\partial x}{\partial t}\right)$ are respectively the acceleration and velocity vectors.

τ represents the time during which the various magnitudes are observed. A total Lagrange formulation is used in this study; thus, we need to determine the variables with respect to an initial configuration taken as a reference.

The vector ${}^{\tau}F_{ext}$ is the excitation load at the time $\tau = t + \Delta t$.

The dimension of the matrix ${}^{\tau}B$ is n x r, ${}^{\tau}B$ corresponds to the position of the actuators in the structure and depends on the coordinates of the nodes at the date τ. Since they are unknown, ${}^{\tau}B$ is approximated by the position of the nodes at the time t. The vector ${}^{\tau}F$ represents the straining forces and depends on the displacements that are themselves unknown. In order to overcome this difficulty, Bathe [Ref 5-39] recommends linearisation of this force, which lead us to formulate ${}^{\tau}F$ as follows:

E 5.79 $${}^{\tau}F = {}^{t}K\Delta x + {}^{t}F$$

Where, ${}^{t}_{0}K$ corresponds to the tangent stiffness matrix and Δx to the displacement increase due to the excitation load ${}^{\tau}F_{ext}$.

From the previous system, Djouadi [Ref 5-38] submits a closed-looped driving force ${}^{\tau}F_c$ such as:

E 5.80 $${}^{\tau}F_c = -R^{-1}\,{}^{\tau}B^{T}\,{}^{\tau}\overline{K}^{-T}\left[Q_1\,{}^{\tau}x + a_4 Q_2\,{}^{\tau}\!\left(\frac{\partial x}{\partial t}\right)\right]$$

with

E 5.81 $${}^{\tau}\overline{K} = a_1 M + a_4 C + {}^{t}K$$

Figure 5.48 represents a cantilever beam composed of four assembled tensegrity units (simplex). The trusses have a length of 1.67 m, 0.02 m of diameter and a modulus of elasticity equal to 200 000 MPa.

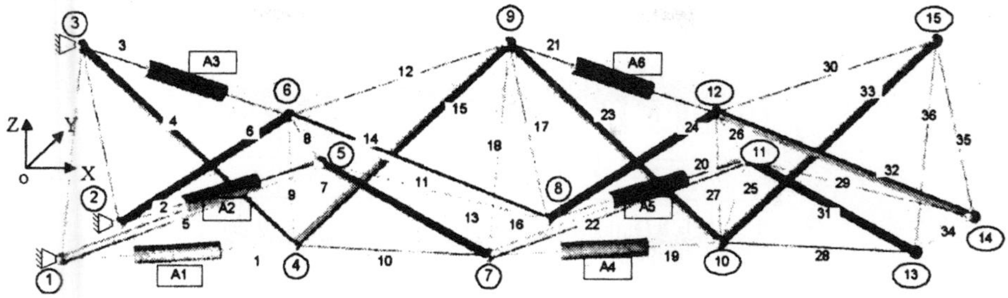

Figure 5.48 *Tensegrity beam*

The characteristics of the cables are given in Table 5.4:

Table 5.4 *Characteristics of cables*

cables Number	Diameters in (m)	Lengths in (m)	Elasticity Modulus in $\times 10^{11}$ Pa	Initial Pre-tensions in $\times 10^{9}$ Pa
1-2-3	0.004	1.43	1	1.7209
10-11-12	0.004	1.43	1	1.7209
19-20-21	0.004	1.43	1	1.7209
28-29-30	0.004	1.43	1	1.7209
7-8-9	0.006	0.57	1	1.3723
25.26-27	0.006	0.57	1	1.3723
16-17-18	0.006	1.67	1	0.7969
34-35.36	0.004	1.67	1	0.4000

The damping of the structure is considered to be negligible. The mass related to all the nodes is equal to 150 Kg. We assume that this assembly simulates the mast of an antenna, which possess an imposed criterion of geometrical form.

In the first simulation, the control is activated with three actuators located on the first unit. In the second simulation, in addition to the three previous actuators three additional ones are used on the third unit. They are all set parallel to the transverse bracing cables. Their function is to ensure the reduction of the nodes' displacements in space.

The materials of the beam components are assumed to have a linear elastic behaviour. The beam is loaded at all nodes by a random type tridimensional external perturbation. It is a normalised signal whose cut off frequency is 50 Hz.

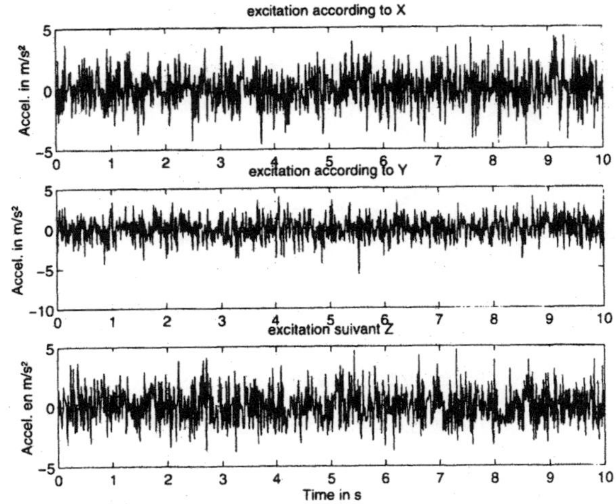

Figure 5.49 *Acceleration in m.s^{-2}, for external excitation according to x, y and z axis*

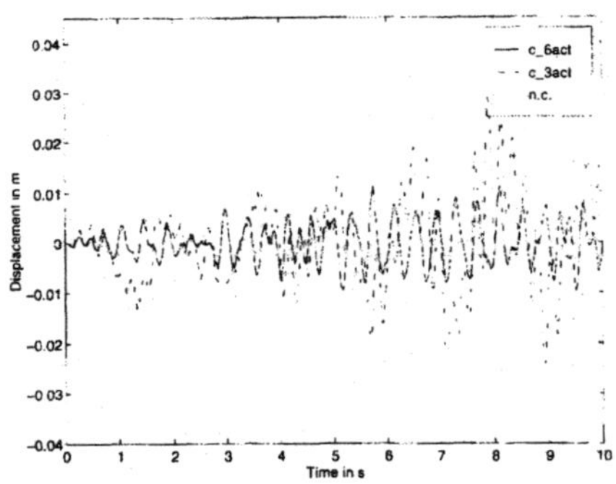

Figure 5.50 *Variation with respect to X-direction, with 6, 3 or 0 actuators*

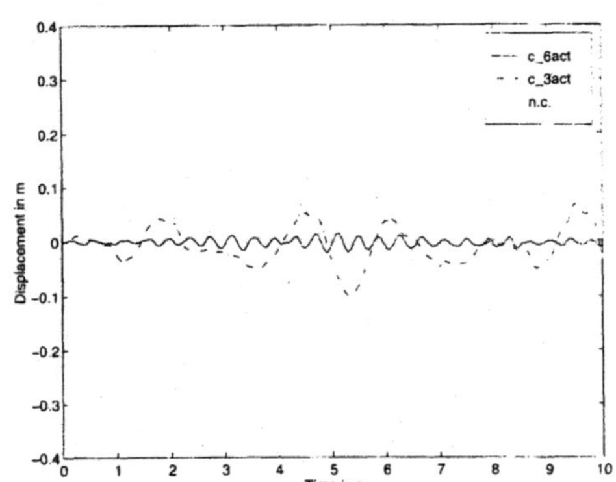

Figure 5.51 *Variation with respect to Y-direction with 6, 3 or 0 actuators*

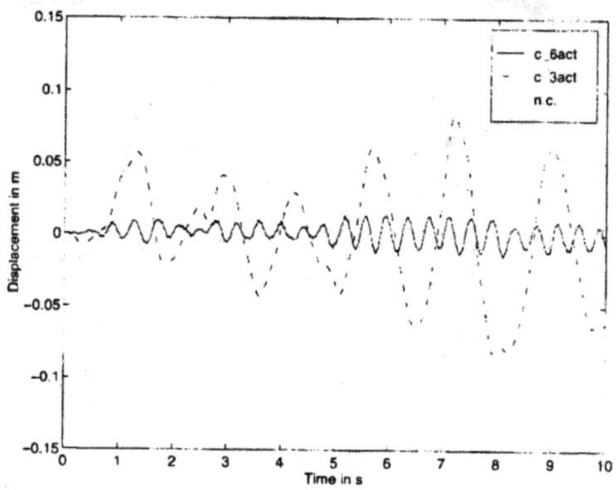

Figure 5.52 *Variation with respect to Z-direction with 6, 3 or 0 actuators*

These figures show the variation with respect to the three directions X, Y and Z. The curves in black result from the use of the actuators. As regard to the control efficiency, a reduction is obtained on the X-axis with an average estimation of 65.05% in the case of the three actuators compared to the uncontrolled displacements and 85.59% in the case of the six actuators. In similar conditions, we have reductions respectively of 85.88% (96.81%) on the Y direction and 51.24 (91.19%) on Z-axis.

5-72.1 Conclusion

The purpose of this presentation was to give but a glimpse of a study on the non-linear control of structures.

We have used a closed looped tridimensional control for this model. We insist moreover that the directions of the actuators do not affect the control, which always remains optimal. The actuators have a favourable effect on the pre-tensions of the cables and consequently contribute to the safety of the structure.

5-8 Conclusion

Knowledge of tensegrity is dependent on basic concepts of mechanics such as self-stress states and mechanisms. A tensegrity system being a system in an equilibrated state, it is important to understand that its design requires a simultaneous monitoring of forms and forces, which are coupled for this type of systems. That is why this chapter was devoted to models and was mainly written with the aim of showing that there are in fact many ways of reaching this dual target. Form-finding processes can be of any nature, but at the end of the day, it is always necessary to check the mechanical properties of the result.

Mechanical behaviour can then be studied with current theoretical models, which are available for systems with initial stresses. We merely wanted to give an illustration of the specific features of tensegrity systems, without any objective for a specific mechanical behaviour model, which remains outside of the scope of this work.

References

Ref 5-1 Vassart N., Recherche de forme et stabilité des systèmes réticulés autocontraints, Thèse de doctorat, Université des Sciences et Techniques du Languedoc, Montpellier, 1997.

Ref 5-2 Micheletti A., "On the cinematics of tensegrity towers", Colloque Lagrange, Rome, 2001.

Ref 5-3 Skelton R.E., Pinaud J.P., Mingori D.L., "Dynamics of the shell class of tensegrity structures", *Journal of the Franklin Institute*, 338/2-3., 2001, pp. 255–320.

Ref 5-4 Smaili A., Fuller Snelson Emmerich, Mémoire de DEA, Université Montpellier II, 2001.

Ref 5-5 Foucher O., Polyèdres et Tenségrité, Mémoire de DEA, Université Montpellier II, 2001.

Ref 5-6 Sultan C., Corless M., Skelton R.E., "The prestressability problem of tensegrity structures, some analytical solutions", to appear in *International Journal of Solids and Structures*, 2001.

Ref 5-7 Motro R., Topologie des structures discrètes. Incidence sur leur comportement mécanique. Autotendant icosaédrique. Note de recherche interne 74, Laboratoire de Génie Civil, Université des Sciences, Montpellier, 1973.

Ref 5-8 Motro R., Equilibre de l'autotendant à une nappe. Note de recherche interne 141, Laboratoire de Génie Civil, Université des Sciences, Montpellier, 1980.

Ref 5-9 Belkacem S., Recherche de forme par relaxation dynamique de systèmes réticulés spatiaux autocontraints, Thèse de Docteur Ingénieur, Université Paul Sabatier, Toulouse, 1987.

Ref 5-10 Linkwitz K., Sheck, H.J., "Einige Bemerkungen von vorgespannten Seilnetzkonstruktionen". Ingenieur-Archiv 40, Springer Verlag, 1971, pp. 145–158.

Ref 5-11 Liu Y., Motro R., "Shape analysis and internal forces in unstable structures", Proceedings of the IASS International Symposium, 1995, "Spatial Structures: Heritage, Present and Future", ed. by G.C. Giuliani, SGE pub., Padovia, Vol. 2, 1995, pp. 819–826.

Ref 5-12 Quirant J., Systèmes de Tenségrité et autocontrainte: qualification, sensibilité et incidence sur le comportement. Thèse de doctorat, Université des Sciences et Techniques du Languedoc, Montpellier, 2000.

Ref 5-13 Pugh A., An introduction to tensegrity, University of California Press, Berkeley, 1976.

Ref 5-14 Williamson D., Skelton R.E., "A general class of tensegrity systems: Equilibrium analysis", in Engineering Mechanics for the 21st Century, SCE Conference, La Jolla, 1998.

Ref 5-15 Sultan C., Modeling, design, and control of tensegrity structures with applications, Ph.D. Dissertation, Purdue University, School of Aeronautics and Astronautics, West Lafayette, 1999.

Ref 5-16 Sultan C., Corless M., Skelton R.E., "Tensegrity flight simulator", *Journal of Guidance, Control, and Dynamics*, 23(6), 2000, pp. 1055–1064.

Ref 5-17 Vassart N., Laporte R., Motro R., "Determination of mechanisms's order for cinematically and statically indeterminate systems", *International Journal of Solids and Structures*, Vol. 37, 2000, pp. 3807–3839.

Ref 5-18 De Jager B., Skelton R.E., "Optimizing stiffness properties of tensegrity structures", in Proceedings of International Mechanical Engineering Congress and Exposition, Vol. 3330, New York, 2001.

Ref 5-19 Kebiche K., Kazi-Aoual M.N., Motro R., "Geometrical non-linear analysis of tensegrity systems", *Journal of Engineering Structures*, 21, 1999, pp. 684–876.

Ref 5-20 Kebiche K., Etude en non linéarités géométriques et homogénéisation des systèmes réticulés spatiaux autocontraints, Thèse de doctorat, Université Montpellier II, 1998.

Ref 5-21 Quirant J., Kébiche K., Kazi-Aoual M.N., "Etude des systèmes de tenségrité", Revue Française de Génie Civil, Vol. 4, 2000.

Ref 5-22 Quirant J., Kazi Aoual N., Motro R., "Tensegrity Systems: Self-stress States and Sensitivity", in Proceedings IASS-IACM 2000, Fourth International Colloquium on Computation of Shell and Spatial Structure, Chania Crete, 2000, pp. 170–171.

Ref 5-23 Yao J.T.P., "Concept of structural control", *Journal of the Structural Division, ASCE*, Vol. 98, N° ST7, Proc. Paper 9048, July 1972, pp. 1569–1570.

Ref 5-24 Abdel-Rohman M., Leipholz H., "Active Control of Tall Buildings", *Journal of the Structural Engineering*, Vol. 109, N° 3, March., 1983.

Ref 5-25 Abdel-Rohman M., Leipholz H., "Active Control of Large Structures", *Journal of the Engineering Mechanics Division, ASCE*, Vol. 108, N° EM5, October 1982.

Ref 5-26 Abdel-Rohman M., Leipholz H., "Active Control of Flexible Structures Dynamic Analysis", *Journal of the Structural Division, ASCE*, Vol. 104, N° ST8, Proc. Paper 13964, August 1978, pp. 1251–1266.

Ref 5-27 Yang J.N., Li Z., Liu S.C., Yang J.N., "Control of Hysteretic System Using Velocity and Acceleration Feedbacks", *Journal of Engineering Mechanics*, Vol. 118, N° 11, 1992.

Ref 5-28 Yang J.N., Li Z., and Liu S.C., "Stable Controllers for Instantaneous Optimal Control", *Journal of Engineering Mechanics*, Vol. 118, N° 8, 1992.

Ref 5-29 Yang J.N., Danielians A., Liu S.C., "A seismic Hybrid Control Systems for Building Structures", *Journal of Engineering Mechanics*, Vol. 117, N° 4, 1991.

Ref 5-30 Yang J.N., Akbarpour, Askar G., "Effect of Time Delay on Control of Seismic-Excited Buildings", *Journal of Structural Engineering*, Vol. 116, N° 10, 1990.

Ref 5-31 Chung L.L., Lin R.C., Soong T.T., "Experimental Study of Active Control for MDOF Seismic Structures", *Journal of Engineering Mechanics*, Vol. 115, N° 8, 1989.

Ref 5-32 Rofooei F.R., Tadjbakhsh I.G., "Optimal Control of Structures with Acceleration, Velocity, and Displacement Feedback", *Journal of Engineering Mechanics*, Vol. 119, N° 10, 1993.

Ref 5-33 Chang C.C., Yang H.T.Y., "Instantaneous Optimal Control of Building Frames", *Journal of Engineering Mechanics*, Vol. 120, N° 4, 1992.

Ref 5-34 Reinhorn A.M., Manolis G.D., Wen C.Y., "Active Control of Inelastic Structures", *Journal of Engineering Mechanics*, Vol. 113, N° 4, 1987.

Ref 5-35 Soong T.T., "Active Structural Control in Civil Engineering", *Engineering Structures*, Vol. 10, 1988.

Ref 5-36 Cha J.Z., Pitarresi J.M., Soong T.T., "Optimal Design Procedures for Active Structures", *Journal of Engineering Mechanics*, Vol. 114, N° 12, 1988.

Ref 5-37 Djouadi S., Le contrôle des structures et les systèmes de tenségrité, Thèse de doctorat, Université des Sciences et Techniques du Languedoc, Montpellier, 1998.

Ref 5-38 Djouadi. S., Motro R., Pons J.C., Crosnier B., "Active control of tensegrity systems", *Journal of Aerospace Engineering*, Vol. 11, N° 2, ASCE, ISSN 0893; Paper n° 16880, 1998, pp. 37–44.

Ref 5-39 Bathe K.J., Ramm E., Wilson E., "Finite Element Formulations for Large Deformation Dynamic Analysis", *International Journal for Numerical Methods in Engineering*, Vol. 9, 1975, pp. 353–386.

6

Foldable Tensegrities

6-1 Introduction

Folding tensegrity systems has been one of our research topics for over a decade. This work began with some initial models [Ref 6-1] and now provides very interesting results so far as numerical models are developed; with introduction of folding processes the field of application of tensegrity systems has opened up considerably. Even the conquest of space can benefit from their characteristics. This is the reason why both practical work and theoretical developments are presented in this chapter.

6-2 Folding principle

6-2.1 Folding tensegrities: a new principle

Folding is required to reduce the volume of objects in space. This operation allows for the transportation and storage of folded objects. Needless to say, the use of folding systems has greatly evolved since the nomadic lives led by primitive man. Nowadays, it covers a wide range of applications, from the simple fisherman's chair to architectural projects and satellite components.

This chapter deals with the folding of tensegrity systems. These systems have intrinsic properties that open up new ways to folding. On the basis of previously described properties of tensegrity systems, it appears that they can be folded and unfolded (and rigidified) by changing the element lengths. Length changes can be applied to both struts and cables. Many different possibilities can be explored, depending on the designer's choice. This is quite different from the classical principle of "scissors", or pantograph structures, which have mainly been developed for spatial structures. In this last case, finite mechanisms are dependent on the existence of many hinges in the structure, and to rigidify the structure one sometimes needs to insert extra elements in the deployed state.

6-21.1 Self-stress and mechanisms

Two concepts are closely related in tensegrity: self-stress and mechanisms. The self-stress notion, which is specific to tensegrity systems, is such that they acquire their rigidity by the stabilisation of infinitesimal mechanisms that exist in the equilibrium geometry. This corresponds, for instance, to the maximum value of a ratio r = s/c, previously described, in the case of three-strut regular tensegrity systems. Nevertheless, for lower values of this ratio "r", one or more finite mechanisms appear and the corresponding systems can be folded by activating these finite mechanisms. Finite mechanisms are special kinds of infinitesimal mechanisms, of infinite order and hence associated strains are equal to zero [Ref 6-2].

6-21.2 Stages of folding

Folding a system consists of reducing its volume. Foldable systems have, therefore, the ability to move from an unfolded configuration occupying a large volume, to a folded configuration having a smaller volume and vice-versa. The process has to be made through a series of intermediate geometrical configurations, without any change to the connections between the different elements. This is what differentiates a foldable system from a system that can be dismantled.

The design of foldable tensegrity system goes through a series of stages that summarise the methodology we adopted. These stages are:

- Mechanism creation.
- Verification of the system compatibility during the process of folding/unfolding.
- Stabilisation and stiffening of the system.

6-212.1 Mechanisms creation

Folding a system requires the introduction of an instability [Ref 6-3], or, more precisely, the introduction of finite mechanisms that will make it possible to transform the system's shape. The creation of mechanisms is usually obtained by suppressing some connections; the choice depends on the "folding mode", i.e. the manner in which the system will be folded.

One can also introduce mechanisms by changing the length of some connections (cables or struts). For classic systems, the procedure is to lengthen cables; the connections to be lengthened are linked in circuits. To design this process, Kwan and Pellegrino ([Ref 6-4] and [Ref 6-5]) introduced the notion of active and passive cables. Active cables are cables that run across the system and, when relaxed, allow the simultaneous lengthening of the cables needed to create mechanisms. Pulling these cables causes the structure to unfold. Passive cables impose a limit on the unfolding of the system.

The mechanisms are activated by applying actions in the required direction of folding, to lead the system from the unfolded to the folded geometry.

6-212.2 Geometrical compatibility

Creating mechanisms gives, necessarily, one or more degrees of freedom to some nodes. The position of these nodes in space depends on the direction of the actions which are applied to the structure. Since there is an infinity of directions, there is a potential infinity of node locations as a function of time. This location depends on:

- The number of degrees of freedom of the nodes.
- The type (strut or cable) of connections with other nodes.
- The degree of freedom of the nodes to which the specified node is linked.

When creating mechanisms, we can consider that there is an infinity of "trajectories", that lead the system from an unfolded geometry to an infinity of more or less folded geometries. Among all these trajectories there are some that do not lead to a folded form of the system, or that are not compatible with its geometry. This can cause undesirable deformations. To avoid this, it is necessary to restrict the nodes' displacements to one trajectory (meaning, a unique combination of trajectories for all the system); this is achieved by reducing the number of their degrees of freedom by fixing their displacements to specific planes.

6-212.3 Stabilisation of the system

The stabilisation of a tensegrity system is obtained, in a given geometry, by the elimination of all finite mechanisms. This is achieved when equilibrium geometry is reached. Then, increasing the level of the self-stress increases the stiffness of the system. This is a key advantage since it is not necessary to introduce new elements in the system, as it was not necessary to suppress elements to create finite mechanisms.

6-21.3 Folding modes

6-213.1 Basic idea

Mechanism creation and system stabilisation are the two issues that need to be addressed. Geometrical compatibility depends upon the chosen folding strategy, closely related to the system studied (mast, grid...).

The so-called equilibrium geometry for a tensegrity system is related to both the relational structure (list of elements and assembly mode) and the length of the elements. For simple cases there is one length for the cables "c" and one for the struts "s". The corresponding tensegrity systems are completely defined by their relational structure and the value of the ratio s/c. Form-finding procedures can provide a specific value r_0 for this ratio. With a higher value the system is in a self-stress state, which can stabilise any remaining infinitesimal mechanisms. For a lower value some finite mechanisms occur. The value of s/c can be derived either by changing "s" or "c". Similar remarks can be made for irregular systems in terms of multiple r_i ratios between different couples of values s_i, c_i.

6-213.2 Folding modes

Length modifications can affect either the struts or the cables; the designer can explore different possibilities. A first classification could consider three main modes:

- "Strut mode", when only strut lengths are modified.
- "Cable mode", when only cable lengths are modified.
- "Mixed mode", when both element lengths are modified.

For each mode, regular or irregular modifications can be made, depending on whether these modifications are identical or not for all the corresponding elements.

The choice between these three modes is closely related to the project, its size, its required characteristics, and also to the folding process requirements. Some preliminary remarks are made in the following sections.

6-213.3 Folding problems

Folding tensegrity systems uses a new principle – and new problems are associated with it. The folding process is closely related to node trajectories, and generally uniqueness of these trajectories is required. Once the folded and unfolded geometries are known, displacement restrictions have to be imposed on some nodes not only to guarantee the uniqueness of trajectory described by the nodes, but also for a matter of compatibility with the folding of adjacent modules in the case of assemblies of several modules.

Two kinds of difficulties are to be underlined.

The first is encountered when two struts come into contact during their own displacement; specific numerical studies have to be carried out, since in this case the contact point is not known *a priori* (slipping has to be taken into account), and it is also associated with a relative rotation between the two struts. This kind of problem is of a mechanical nature, since its solution requires a numerical analysis (which can be very complex in the case of multiple contact points).

The second difficulty is of a design nature. A mechanical behaviour study of tensegrity systems shows that, according to the number of self-stress states, it is not always necessary to act, for instance, on all the cables when using "cable mode". The choice of one cable or a set of cables leads to the notion of "active cable", which has been used by Kwan and Pellegrino for other systems. An example is presented later in this chapter.

6-2.2 Strut mode folding

"Strut mode" folding has been used by other researchers such as Hanaor [Ref 6-6] and Furuya [Ref 6-7] for three strut modules and their assemblies in the form of grid and masts. The introduction of finite mechanisms results from shortening the struts.

Mechanisms: creation and activation. Using telescopic bars (which can be shortened by means of a bolt) creates finite mechanisms. A nut and a bolt fixed on the strut of greater diameter allow the locking of the strut to a given length (Figure 6.28). Mechanisms appear when the bolt is unlocked. Once folded, the module is reduced to a bundle which is approximately the length of the collapsed strut. The six-strut module is used here as an example (Figures 6.1 and 6.2).

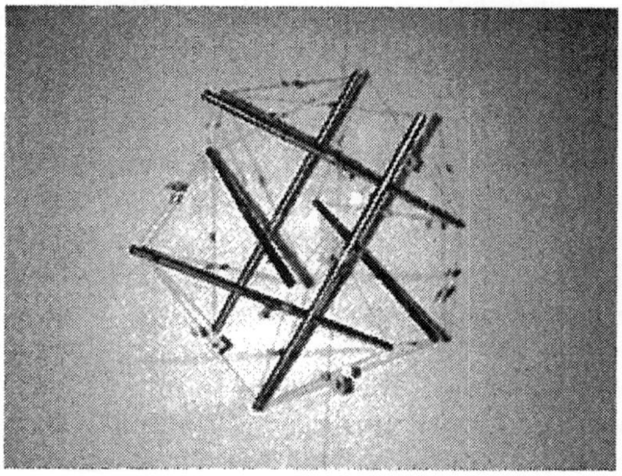

Figure 6.1 *Six-strut tensegrity system: deployed state*

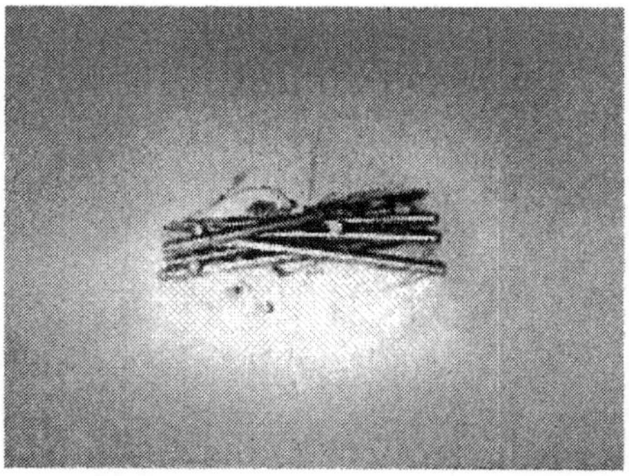

Figure 6.2 *Six-strut tensegrity system: folded state*

The main disadvantage of this mode is the unfolding of the system. Indeed, the folded system appears as a shapeless bundle of telescopic-struts and cables; the first attempt to lengthen the struts is often opposed by the resistance of an inextricable tangle of cables.

It is therefore necessary, despite appearances, to arrange the struts in the position required for them to start unfolding. It is obvious that in order to succeed it is necessary to be well aware of the composition of the system to unfold.

6-2.3 Cable mode folding

6-23.1 Introduction

The main property of this folding mode is that an appropriate choice of the cables that are lengthened to introduce finite mechanisms may lead to an elegant and easy folding process. The choice of this cable determines the folded geometry. Our studies on physical models led us to proceed as Pellegrino did with an "active cable", which is progressively lengthened, thus allowing a better control of the folding process.

We took as an example the same module that was used for "strut mode" folding. The early studies on a physical model were useful to define a folding strategy, based on the fact that two opposite triangular faces of the module could move one relative to the other by a simple translation without any relative rotation. During the motion the triangles remain in two parallel planes. This property allows the folding of the module until the two triangles become coplanar, this common plane being either the middle one (and then the corresponding folding is called "bilateral folding") or the plane containing one of the two opposite triangular faces ("unilateral folding" illustrated in Figure 6.3). Moreover, both possibilities are compatible with the chosen folding strategy for an assembly of six strut modules forming a mast.

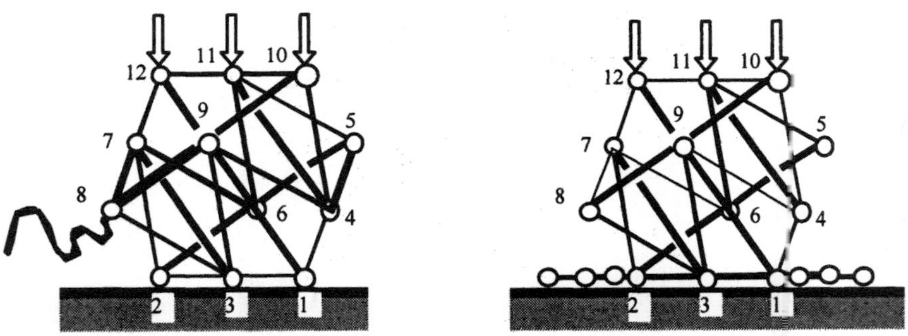

Figure 6.3 *Cable-mode folding of a six-strut tensegrity system*

6-23.2 Mechanisms: creation and activation

<u>6-232.1 Mechanism creation</u>

The choice of a set of cables constituting an "active cable", which could be lengthened (thus introducing finite mechanisms) is governed by several considerations: respect of symmetries, continuity of the cables which have to constitute a circuit going through one and only one of the extremities of each strut and respect of the folding policy as previously defined. Graph theory is very useful at this point. Examining a six-strut system as an assembly of two antisymmetric three-strut systems and six extra cables leads to study the assembly of two similar cable graphs (Figure 6.4). These are graphs of cables for regular three-strut systems. We previously used graph theory [Ref 6-8] and established some results concerning "spherical tensegrity modules". For instance, for these modules the cable graphs have to be planar (no link intersection).

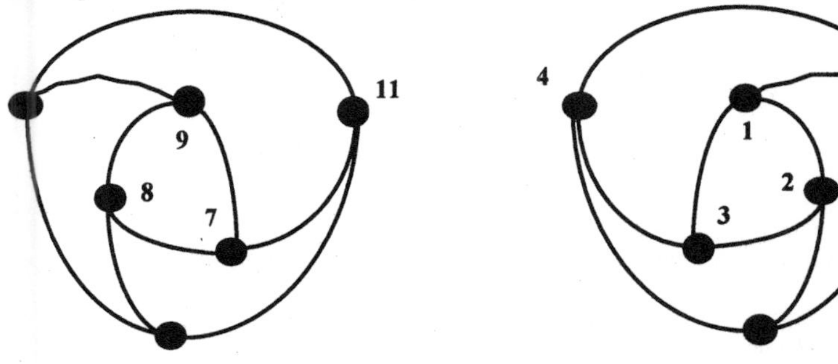

Figure 6.4 *Cable graphs of three-strut tensegrity systems*

Since a six-strut module can be built by the addition of two regular three-strut modules, we superimposed the two graphs and assembled them. Without modifying the number of nodes and links, a whole graph is created, thus putting in evidence a circuit (Figure 6.5). Nodes 4, 5 and 6 are put respectively between nodes 8 and 9, 9 and 7, 7 and 8. Nodes 4, 9, 5, 7, 6, 8, 4 then describe the circuit.

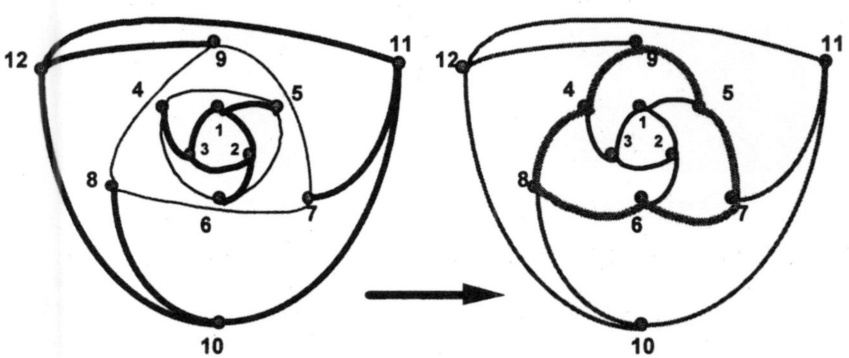

Figure 6.5 *Superimposition of two graphs and circuit emergence*

Such a procedure can be automated, since the problem is to find a circuit in a planar graph, such that each node is at the extremity of a strut (complementary disconnected graph); all struts have one end on this circuit and, of course, the other end is not on the circuit. It is then possible to complete the graph keeping it planar by the addition of six more cables (Figure 6.6). The cable circuit 4, 9, 5, 7, 6, 8, 4 constitutes the active cable required.

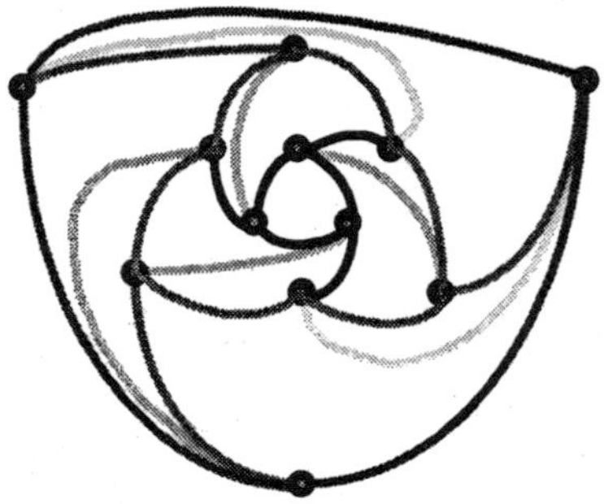

Figure 6.6 *Complete cable-graph of a six-strut tensegrity module*

6-232.2 Mechanism activation

A model has been built with pulleys (Figure 6.18) allowing the active cable to be lengthened to activate the mechanisms (Figure 6.7).

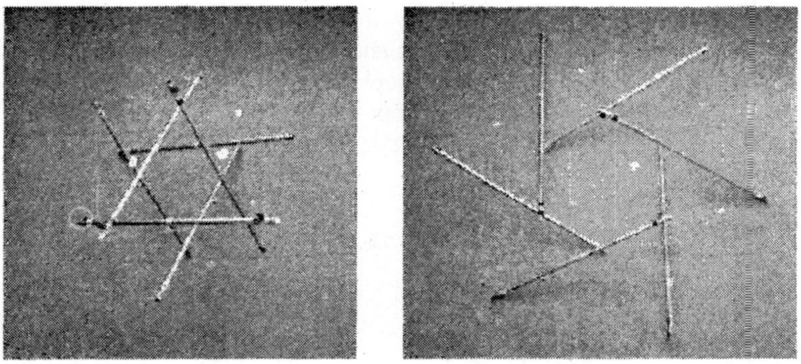

Figure 6.7 *Unfolded and folded six strut tensegrity system*

Relaxing the active cable at one node only introduces the mechanisms. In the physical model, self-weight of the constitutive elements can induce small perturbations in the folding process, which can be numerically modelled as "unilateral" or "bilateral". For each case it is a motion with controlled displacements. Physically, it is more difficult to carry out bilateral folding.

6-23.3 Folding process, deployment and stabilisation

6-233.1 Folding process

Two main remarks could be made on the basis of the physical model test: the first concerns the relative motion of two opposite triangular faces, and the second the non-occurrence of strut contact during the folding process. It is then possible to numerically model node trajectories for both cases – unilateral and bilateral folding. No details are given here on the numerical model that has been developed.

6-233.2 Deployment and stabilisation

Deployment occurs by pulling the active cable (at a single node) until the elimination of all finite mechanisms. By pulling more on the active cable, we introduce self-stress, which stabilises the infinitesimal mechanisms of the structure. Nevertheless, the substitution of six peripheral cables linked to fixed nodes by a single active cable passing through pulleys increases the number of mechanisms, and this must be carefully studied.

6-23.4 Conclusion to "cable mode" fo lding

Folding a tensegrity system according to "cable mode" presents some real advantages even if volume reduction cannot be compared with "strut mode" for which the system can be reduced to a simple bundle:

- A single control point is needed for mechanism activation, folding, deployment, and stiffening.
- Cable entanglements are avoided by the use of the active cable.
- Practical design implementation is easier for this folding mode.

6-2.4 Conclusion

Our main objective was to test the feasibility of different folding principles that make use of the main properties of tensegrity systems. Two folding modes have been identified. "Cable mode" seems to be promising for large systems and could be mixed with a limited use of strut-mode to optimise the folding process. A more detailed design study and numerical modelling is required.

6-3 Foldable modules

6-3.1 Introduction

Before dealing with complex systems it was useful to carefully study the four-strut and the six-strut cells – some classic modules.

6-3.2 Four-strut module

6-32.1 Geometry

This tensegrity module derives geometrically from the half-cuboctahedron.

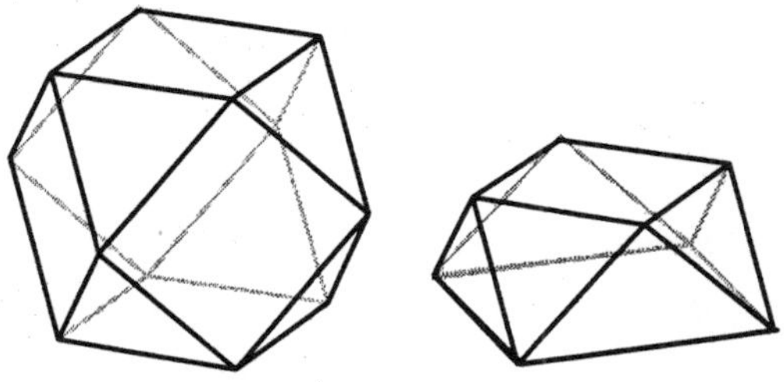

Figure 6.8 *Cuboctahedron and half cuboctahedron*

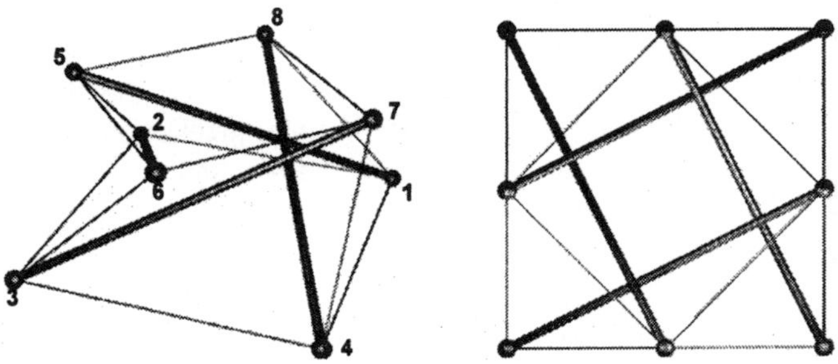

Figure 6.9 *Four-strut tensegrity system*

It has four-fold symmetry and comprises four struts and twelve cables (Figure 6.9). It is not a "regular" system, since its upper cable square layer is inscribed, in plane projection, in the lower cable square layer. This geometry allows aggregating similar modules in order to generate the double layer that will be described in this chapter.

6-32.2 Folding of the four-strut tensegrity system

It is necessary to choose a folding strategy. The one that we chose is compatible with the geometry of the system (module) and with the presumed geometry of an assembly of these modules, to constitute a bigger system. The folding strategy that satisfies these conditions is a projection on the plane containing the nodes 1, 2 and 8 (Figure 6.10). The plane of projection will be referred to as the "folding plane".

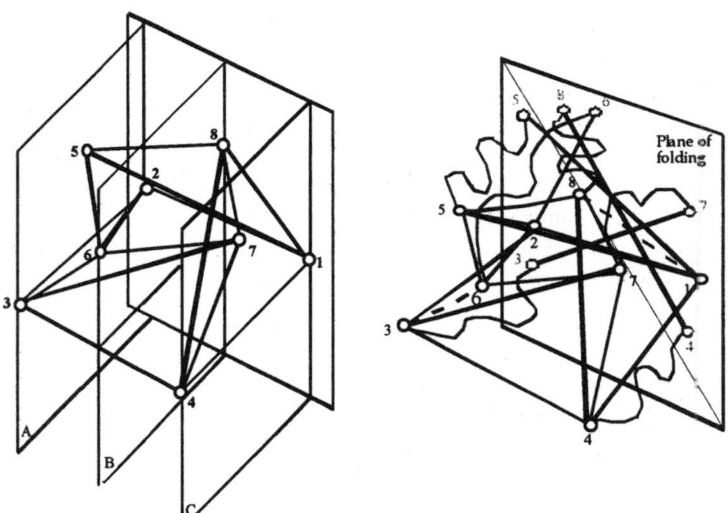

Figure 6.10 *3D view and folding prediction*

The next step is the lengthening of the cable links, so that the module can be folded. The strategy to find which links have to be lengthened is as follows:

- Activation of mechanisms that are orthogonal to the folding plane (Figure 6.11, with the final position in grey lines).
- Lengthening of the minimum number of links required to fold.

The result of this operation is the lengthening of connecting cable 1-8 and its symmetrical 3-6. Figure 6.13 shows the system after activation of mechanisms.

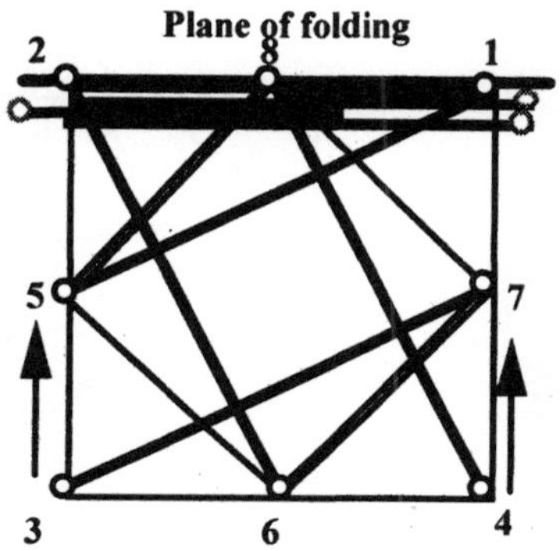

Figure 6.11 *Mechanism activation*

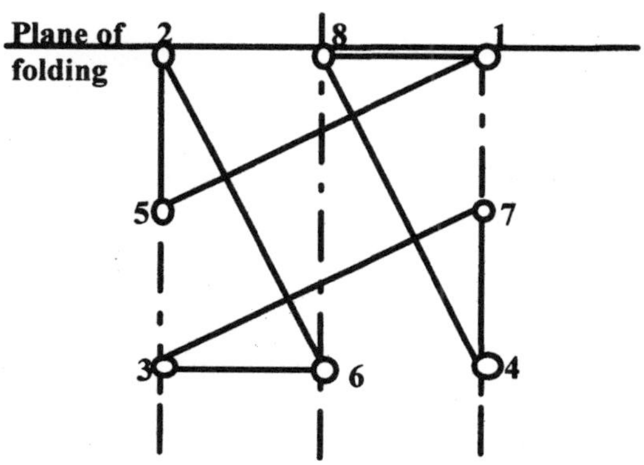

Figure 6.12 *Boundary conditions*

The system in its folded position as indicated in Figure 6.13 (grey shadow) is not the only solution, but one among many. Each of these "folded geometries" stems from the intersection of a combination of trajectories, with the plane of folding. To avoid blockage of the structure, the displacement of some nodes is fixed in particular planes. These planes (A, B and C, Figure 6.10) materialise physical limits of the module and form a "corridor" in which the module can be folded without interference with neighbouring modules (see Figure 6.12).

For this module, several models have been developed and are described next:

- A physical model.

- A geometrical model.
- A numerical model.

These three models do not take into account exactly the same conditions. But even if the results are different – they lead to similar conclusions.

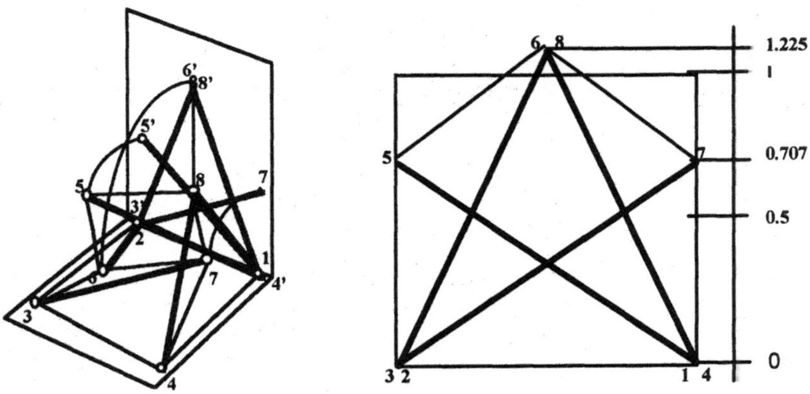

Figure 6.13 *Nodes trajectories and final configuration*

6-3.3 Six-strut module

6-33.1 Geometry

The six-strut module comprises six struts and twenty-four cables (Figure 6.14). This regular tensegrity system is characterised by the ratio $r_0 = 1.67$.

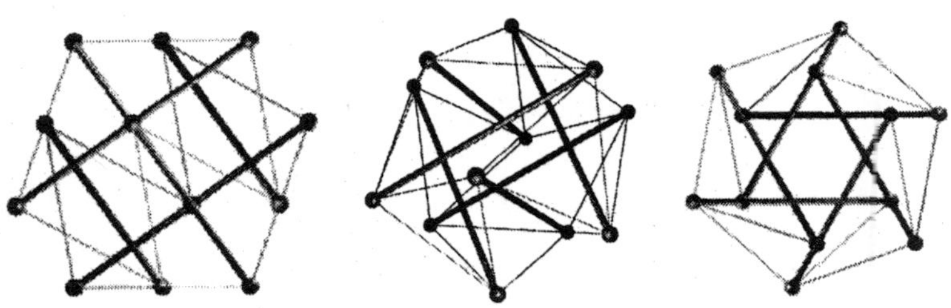

Figure 6.14 *Six-strut tensegrity module*

6-33.2 Strut mode folding

Other researchers (Hanaor, [Ref 6-6], and Furuya [Ref 6-7]) have already investigated the creation of mechanisms by strut shortening. The main advantage of this mode is that it allows a strong reduction of the volume of the system, since it reduces the length of the struts that are the only formally rigid elements in the system.

The physical modelling of folding is made through small-scale models of modules with telescopic struts. Two telescopic tubes constitute each strut. Once folded, the module is reduced to a bundle having approximately the minimal strut length.

To begin with we sought to avoid using this mode, which is technologically complex, but we decided to test it on a six-strut module. This had never been carried out before because of the great associated volume reduction.

6-33.3 Mechanism activation

Considering that the self-weight of the system is neglected, it is sufficient to push on the two parts of the telescopic strut to activate the mechanism so as to shorten it. The reverse operation is required for the unfolding process. The system's scale is very important here. In fact, for small-scale systems the struts can easily be shortened individually by hand. This becomes impossible for large scale and heavy modules and this clearly cannot be the case for major architectural projects. Fluid pressure could be used in such cases, as A. Hanaor demonstrated, but this solution is technologically complex, and also expensive.

6-33.4 Physical model

A physical model has been built to test these manipulations (Bouderbala, [Ref 6-9]). Because the folded geometry is not unique, several solutions can be realised and each applies different constraints on the strut shortening operations. Two of these solutions are described: *bundle type* and *plane type.*

- Bundle type:
  All struts are identically reduced to half of their initial length The module can then be folded in a bundle of shortened telescopic bars and all the cables are slack; volume reduction is maximum.
- Plane type:
  It is not possible to completely map the module onto a plane with "strut" mode. Regular folding can be achieved. The module keeps a little depth and most cables are taut. However, in this case we have lost the main advantage of this mode: for the reduction of volume is not maximal. When handling the model, it is clear that it would be possible to completely flatten the module by acting on the cables (it would then be what we called "mixed mode").

An irregular folding was tested, taking into account the fact that the six-strut module can be considered as an assembly of two three-strut modules with a few extra cables (Figure 6.15).

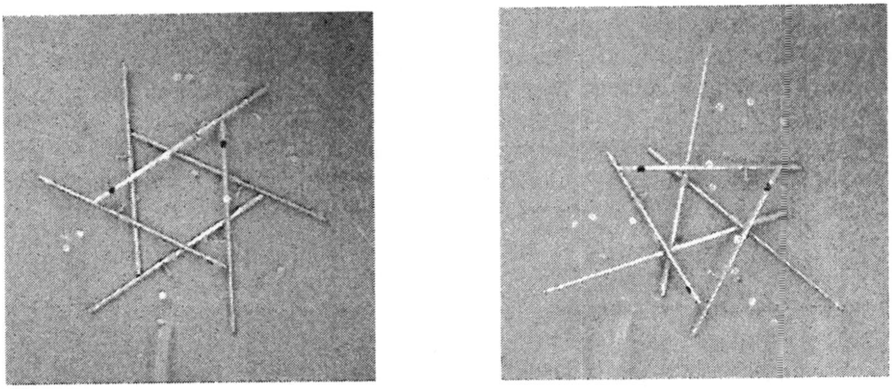

Figure 6.15 *Two possible flattened configurations*

6-33.5 Cable mode folding

The folding of the six-strut module is made by a projection of the module onto the plane containing one of the triangular faces. In our case the module is folded down to the plane containing the nodes 1, 2 and 3. In order to do this, it is necessary to lengthen cable links 4-5, 5-6, 6-7, 7-8, 8-9 and 9-4. These links constitute a circuit that is used advantageously for the creation of mechanisms. Thus, these connecting cables form an active cable passing through nodes 4, 5, 6, 7, 8 and 9 (Figure 6.16). The creation of mechanisms and their activation leads to a folded geometry of the module that, in the case of the 6.struts module, is a unique solution. The lower and upper faces defined respectively by nodes 1, 2, 3 and 10, 11, 12, do not change form nor do they undergo rotation. In fact, in a folded geometry, the upper face (10, 11, 12) overlaps the lower face. Trajectories of the nodes 4, 5, 6, 7, 8 and 9 are close to arches of circle whose centres are nodes 11, 2, 12, 10 and 1 respectively.

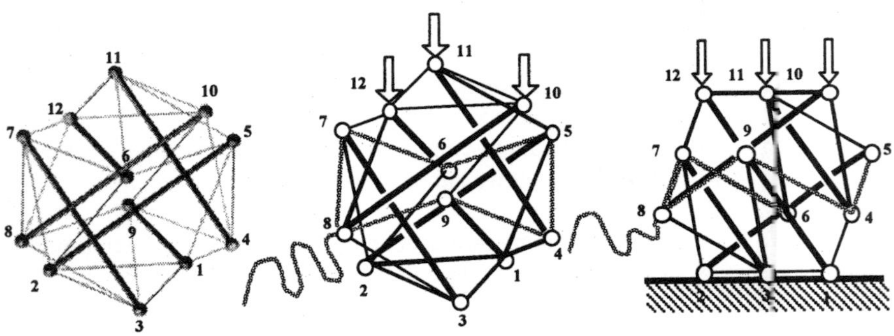

Figure 6.16 *Folding with an active cable*

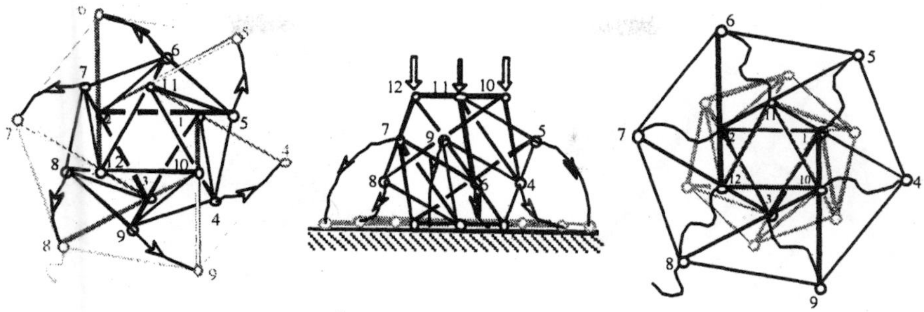

Figure 6.17 *Nodes trajectories*

Figure 6.18 *Detail of the pulley*

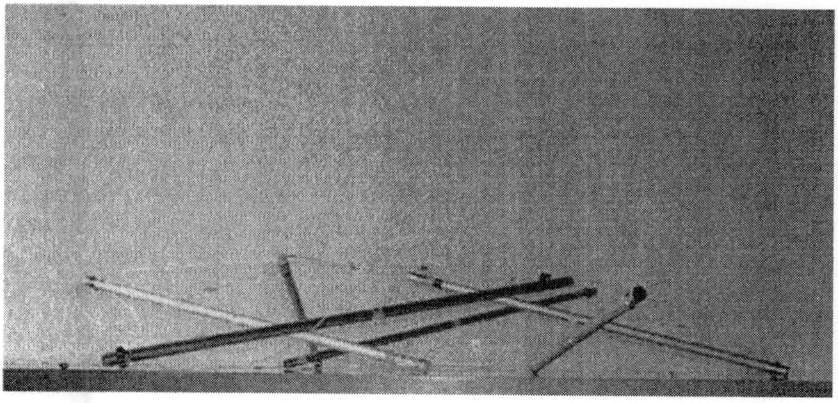

Figure 6.19 *Flattened shape*

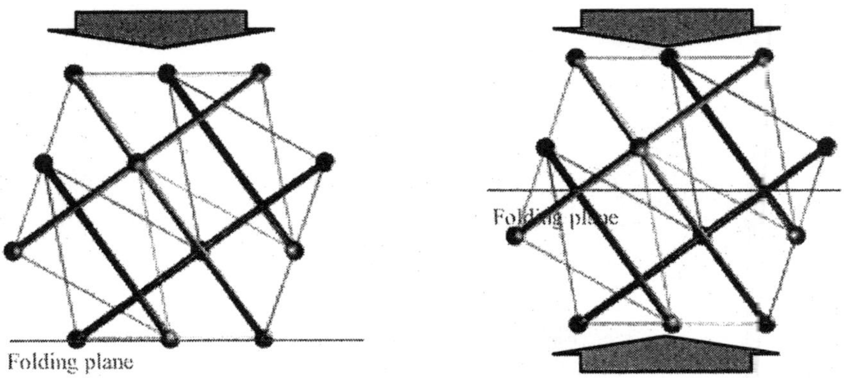

Figure 6.20 *Unilateral and bilateral folding*

The module is unfolded by pulling the active cable until the lengths of the cables are all identical, i.e. until the self-stress geometry.

6-4 Foldable assemblies

6-4.1 Tensegrity mast

There are two ways to combine six-strut modules and to form mast structures. The first is based on sharing a triangular face between two modules; the second is based on sharing a hexagonal face. The first mode of assembly requires the design of a special node joining the struts of the assembled modules. The second mode does not call for the design of a specific node.

Projecting the mast onto a plane will bring about the folding. Active cables control the creation of mechanisms and the unfolding of the mast. If it is required that the surfaces joining the upper and the lower faces of the module do not undergo any deformation or rotation, each module is independent. It therefore follows that the folding/unfolding of the mast will be on a module-by-module basis, and the state of self-stress will also be for each individual module. This result requires an active cable per module (third view in Figure 6.21). If this procedure is not necessary, the active cables of modules will be linked to form a single active cable that will allow the folding/unfolding of the whole mast. The fourth view of Figure 6.21 shows such a case.

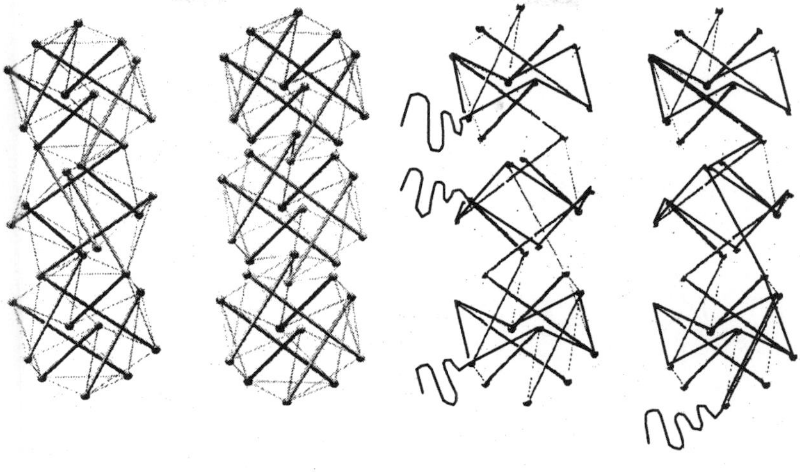

Figure 6.21 *Two mast assemblies and two folding strategies*

A physical model (Figure 6.22, [Ref 6-9]) was created with the folding strategy that has been established with three active cables.

The hypothesis on the pure translation of basis triangles was validated by the physical experience illustrated in the following figures[1].

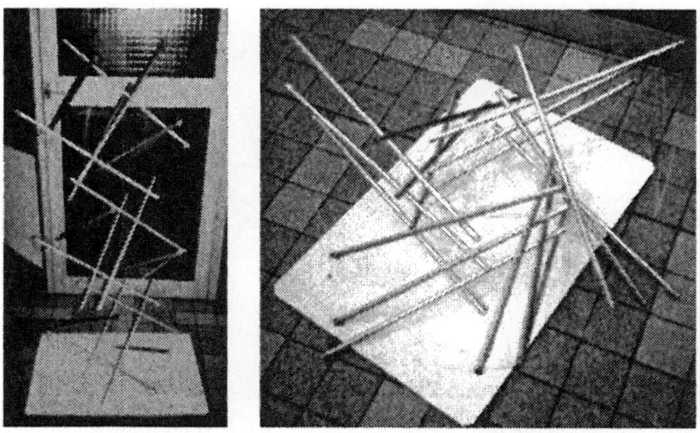

Figure 6.22 *Physical model of a tensegrity mast*

6-4.2 Double-layer grid

The assembly of nine single tensegrity cells gives a double layer grid. We built a prototype in order to test its folding solutions. The choice of a folding process is shown in Figure 6.24. The links to be lengthened in the grid in order to create the required mechanisms are the same as those already identified for the module. These

[1] Cornell Sultan recently confirmed to the author that he had found the same results with a theoretical model.

links in the grid form chains that are used to facilitate the creation of mechanisms and the unfolding of the grid.

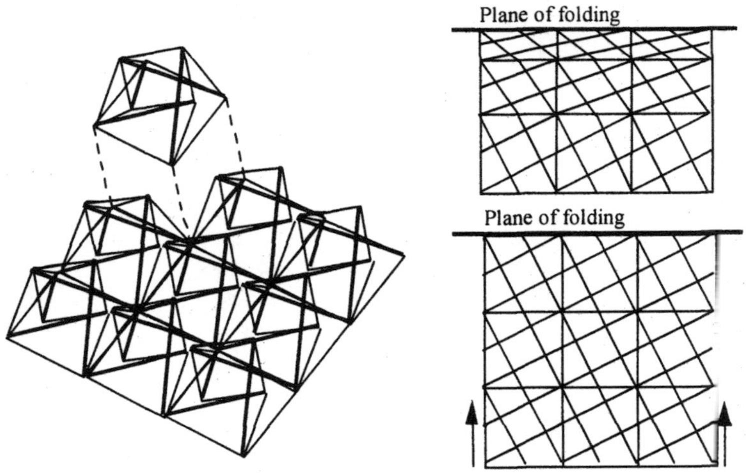

Figure 6.23 *Tensegrity double layer grid folding policy*

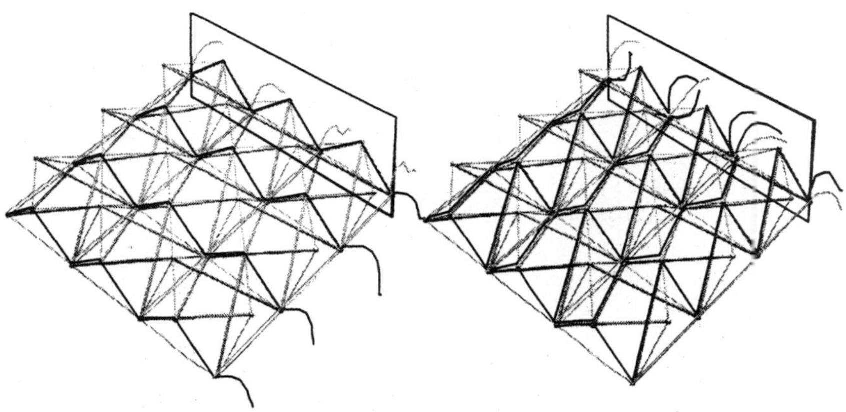

Figure 6.24 *Two folding policies*

There are at least two ways to define chains. The first is simple to build and has been investigated by making a physical model. The second possesses the advantage that the active cables terminate in the folding plane (Figure 6.24).

To activate the mechanisms and to fold the grid starting from the folding plane, we then added a second series of cables. The latter allows folding the grid by simply pulling the nodes towards the folding plane.

The uniqueness of the trajectories and the respect of the geometrical compatibility of the system during the process of folding/ unfolding depend on the care with which the nodes have been designed. A detailed study of the folding/unfolding of the module is required, as the mutual spatial relationship between struts and nodes remains the same at both the module and grid scale. A study based on a numerical model aimed at implementing the design of nodes is being carried out in the Laboratory of Mechanics and Civil Engineering at the University Montpellier II.

The stabilisation and self-stress of the grid are obtained by means of the same active cable.

The two extreme positions are illustrated in Figures 6.25 and 6.26.

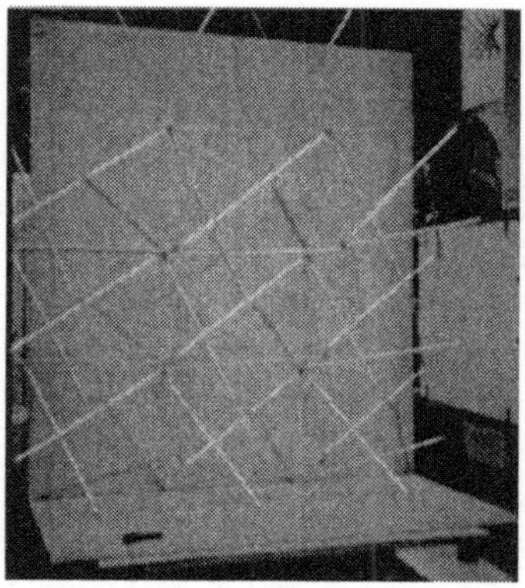

Figure 6.25 *Double layer physical model: deployed state*

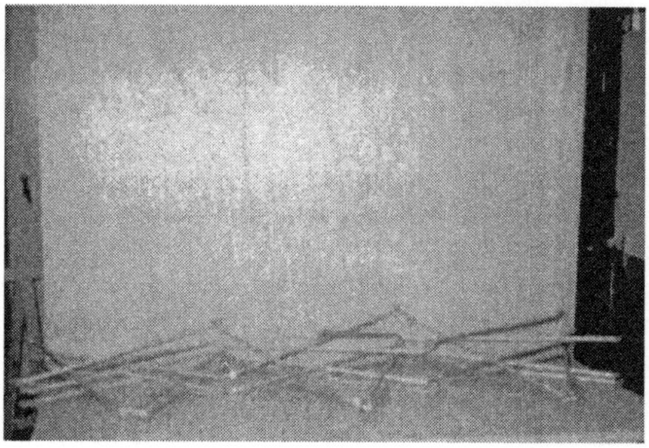

Figure 6.26 *Double layer physical model: folded state*

6-4.3 Conclusion

In conclusion, this work shows that folding tensegrity modules with four and six struts is indeed possible. Physical models for the two modes of creating mechanisms have validated this.

In so far as assemblies are concerned, folding them by shortening the struts was excluded because it highlights the complexity of tensegrity systems without providing any solutions.

On the other hand, folding an assembly by lengthening its cables is very promising. The grid of four-strut modules that has been investigated by means of a physical model fully met our expectations. This concept is currently the object of a numerical model that will allow us to know more precisely the trajectories of the moving nodes, and which will enable us to design the nodes of module assemblies. A physical model of the mast with six-strut modules is now being prepared (Figure 6.22).

We hope in the near future to be able to explore other active cable paths for both the four-strut module grid and the six-strut module mast. We will also attempt to apply a textile or rigid cover over the grid: this will allow the utilisation of systems such as variable geometry architectural roofs.

6-5 Folding design

6-5.1 Introduction

Tensegrity systems are complex, at least from a geometrical point of view. The design of folding tensegrities requires several simultaneous approaches. A large range of problems has to be solved; these problems involve design numerical modelling, mechanical, and technological studies. We have chosen to work according to a particular methodology which associates three different approaches:

- Physical models are essential for design and for examining the process, taking full account of real physical constraints. It allows one to choose a particular folding policy.
- Geometrical modelling can sometimes be useful to describe node trajectories and member displacements. This modelling is based mainly on the use of sphere intersections.
- Numerical modelling is an essential tool to predict the process, but its refinement needs a lot of work to take into account contact, friction problems etc. Graphical simulations are, of course, very useful to visualise the folding process. The theoretical basis for this numerical modelling is presented in this section. Only certain results are presented for comparison with physical and geometrical studies.

6-5.2 Four-strut tensegrity folding

6-52.1 Physical modelling

6-521.1 Geometrical description

In this section, we have developed in detail the case of the four-strut tensegrity module. It is a first step towards a study of the double layer grid. The tensegrity module with four struts derives geometrically from the half-cuboctahedron. It has an axis of symmetry of order four and it comprises four struts and twelve cables (Figure 6.27).

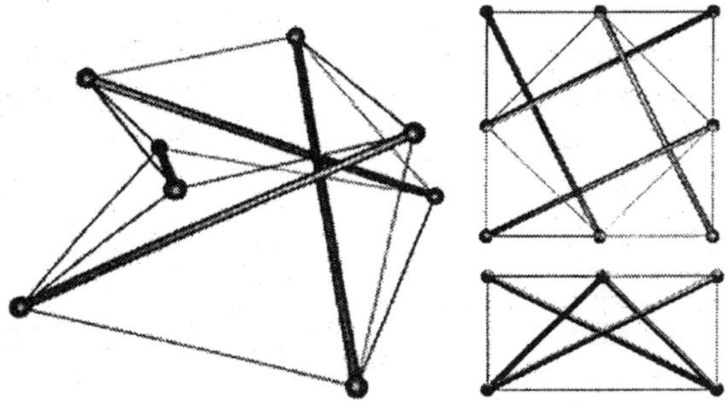

Figure 6.27 *Four-strut tensegrity system: perspective, plan and elevation views*

6-521.2 Folding mode

The folding of the four-strut module by shortening the struts has been achieved by means of telescopic cylindrical struts. The physical model shows that this way of creating mechanisms gives the greatest reduction of the system volume. But unfolding is very difficult. And in order to be carried out correctly it calls for the system to be well known by users. The work of Hanaor [Ref 6-6] in this field shows the difficulty of applying such a folding mode to an assembly of modules.

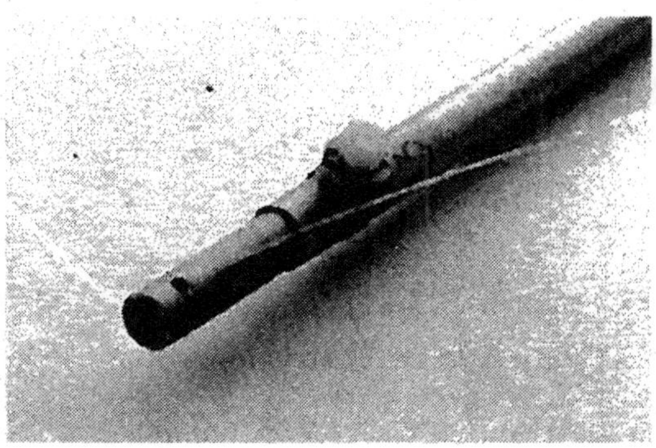

Figure 6.28 *Telescopic strut*

Let us now focus our attention on the second mode of creating mechanisms (i.e. by lengthening the cables). Folding by lengthening the cables of the four-strut module has been achieved by means of a small-scale model consisting of aluminium tubes linked by nylon cables.

6-521.3 Folding policy

The folding policy (the manner in which the module is folded) in this case is a flattening of the module in the plane containing the nodes 1, 2 and 8. This plane has been called the "folding plane" (see Figures 6.10 and 6.11).

The final aim of the study being to fold an assembly of modules, the displacement of some nodes has been fixed in specific planes representing the physical edges of the module. These arrangements make the module fold in a corridor, thus preventing all risk of deformation of the constitutive elements of the module due to interference between modules during the folding process of the double layer grid that is to be formed.

Consequently, the displacement of the lower layer nodes has been fixed in the horizontal plane. Concerning nodes 5, 7 and 8 (node 8 being in the folding plane), since there is no practical way of fixing them to lie in vertical planes, we allowed them to slide freely on flat surfaces, which are materialised by these planes. Nodes 3 and 4 being located at the intersection between two planes (horizontal and vertical planes), slide over a groove materialising the intersection line. Nodes 1 and 2 are at the intersection of three planes and hence are fixed. Node 6 is totally free.

6-521.4 Operating mode

Mechanisms were introduced by replacing the bracing cables (i.e. cables 1-8, 2-5, 3-6 and 4-7) by rubber bands. This enables folding of a module in the folding plane under the action of small forces.

Experiments show that it is not necessary to replace all the bracing cables (four cables) but only two of them, cables 1-8 and 3-6. To activate the mechanisms and initialise folding, we used the self-weight of the struts.

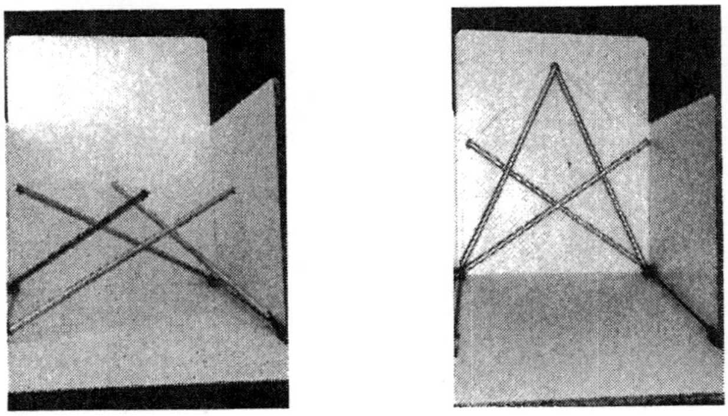

Figure 6.29 *Physical model: unfolded and folded configurations (elevation)*

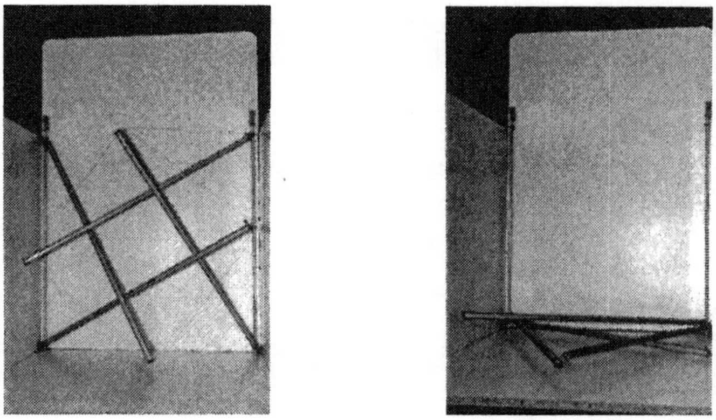

Figure 6.30 *Physical model: unfolded and folded configurations (upper view)*

6-52.2 Geometrical modelling

6-522.1 Geometrical approach

Designing the node of a foldable tensegrity grid requires that its trajectories (and hence the relative motion of struts and nodes) be well known.

We decided, in order to more precisely define the process, to fix the displacement of all the nodes to specific planes, as was the case in respect of the physical model. But when it came to the geometrical model we fixed the displacements of nodes 6 (totally free in the physical model) and 8 in a median plane B (see Figure 6.11).

The trajectories of the nodes are determined afterwards by the geometrical intersection between the "geometric loci" (the set of all the possible positions of the node in space) of the nodes and the displacement planes of these nodes.

Consequently, the plane containing nodes 1, 2 and 8, being the folding plane (F plane), and the plane containing nodes 1, 2, 3, and 4 being the horizontal one (H plane), the resulting displacements are as follows:

- The trajectory of node 1 is at the intersection of three planes F, H and C. Node 1 is fixed.
- The trajectory of node 2 is the intersection of the planes F, H and A. Node 2 is fixed.
- The trajectory of node 3 results from the intersection of the planes A and H. Node 3 slides along the intersection line.
- Node 4 slides along the intersection line of the planes H and C.
- Node 8 slides along the line resulting from the intersection of the planes F and B.
- The case of nodes 5 and 6 is different; these nodes are linked to fixed nodes by means of struts. The geometric locations of these nodes are thus spheres. The

intersection of the geometric constraints on the nodes defines two circles. Hence the trajectories of nodes 5 and 6 are circular arcs.

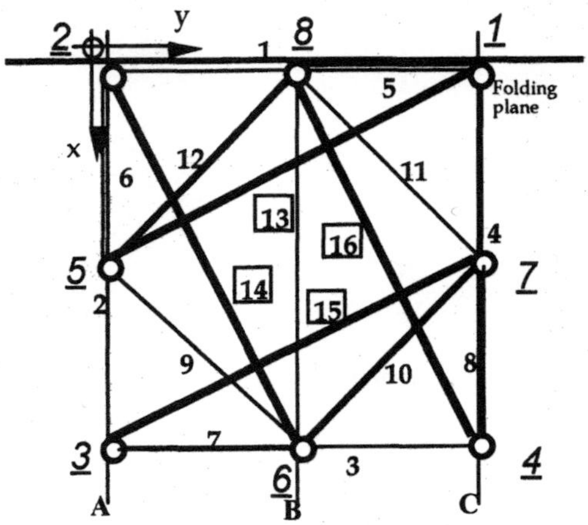

Figure 6.31 *Numbering of nodes (italic, underlined) and members*

Let us take the example of node 5. It is linked to node 1 by means of element 13 (strut). Node 1 being fixed and element 13 being a strut the locus of node 5 (having two degrees of freedom) is a sphere, with its centre at node 1 and the length of element 13 as radius.

The co-ordinates of the centre being 1(0, m, 0); the mathematical expression of the sphere $S_{13.1}$is (x,y and z being the node coordinates):

E 6.1 $$S_{131}: x^2 + (y - m)^2 + z^2 = L_b^2 \quad \text{with} \quad L_b = L_{13}$$

with L_b, length of the strut, and m edge length of the module basis. The intersection between sphere $S_{13.1}$ and the displacement plane A: $y = 0$ gives a circle, which is the trajectory of node 5:

E 6.2 $$A \cap S_{131} = C_{131}$$

E 6.3 $$C_{131}: x^2 + z^2 = L_b^2 - m^2$$

The case of node 7 is different. It is connected to node 3 by means of element 15 (strut) but in this case node 3 slides along a line. The trajectory of the node 7 cannot easily be defined.

The easiest way to give a convenient trajectory to node 7 is to consider the future assembly of the module with other modules. In the grid (planar assembly of modules) node 7 is linked to node 5 of the neighbouring module. Consequently, they have the same trajectory. It has been verified that the displacement of node 7 according to the trajectory of node 5 trajectory involves deformations of the constitutive elements. The resulting trajectories are shown in the following illustrations:

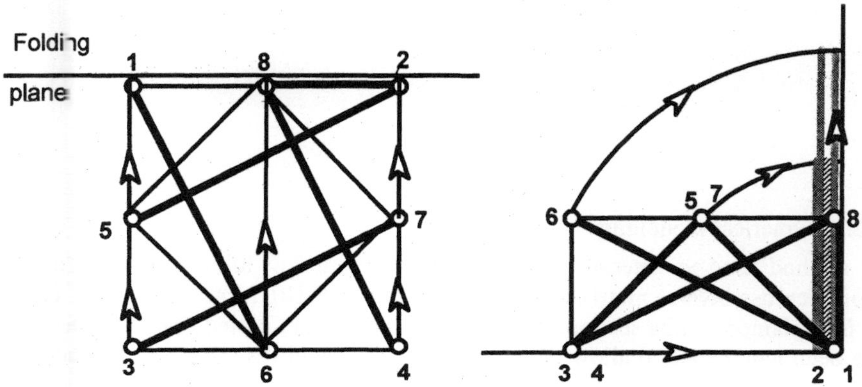

Figure 6.32 *Node trajectories*

The other possible mode is to consider that the displacement of node 7 is due to the traction applied by cable 7-8. In this case the trajectory of node 7 is, at every moment, the result of the intersection between the circle $C_{15.3}$ and the circle $C_{11.8}$:

- The circle $C_{15.3}$ being the intersection of the sphere $S_{15.3}$ (node 3 is the centre and strut 15 of length L_b materialise the radius) and the displacement plane C.
- The circle $C_{11.8}$ being the intersection of the sphere $S_{11.8}$ (node 8 is the centre and cable 11, joining the nodes 7 and 8, is the radius) and the C plane.

<u>6-522.2 Graphical simulation</u>

We carried out a graphical simulation of the folding according to the above geometrical model[2]. This software is able to draw elements between points with given co-ordinates. The trajectories of the nodes were implemented in the software with steps of displacement for nodes 3 and 4.

The software produces for each step a rendered image of the module during the process of folding. The images produced were then converted into moving images, which can be viewed using Quicktime®. Here are some screens taken from such images.

[2] This was carried out with a software known as "µ3D" and developed at the Montpellier School of Architecture (EALR) by Alain Marty.

Figure 6.33 *Motion simulation*

6-52.3 Numerical modelling

The method used to determine different positions during folding is based on the Moore Penrose inverse matrix method (Hangai, [Ref 6-10], Motro *et al.*, [Ref 6-11]) and is explained in this section. The two cables drawn in dotted lines (Figure 6.34) have been lengthened to allow the activation of mechanisms (with external actions) and then folding.

Nodes 3 and 4 are able to slide along the **y** direction, and folding is achieved on the vertical plane containing the three nodes 1, 2 and 8 according to the physical model.

Figure 6.34 shows the tensegrity module in its folded state. Two main points can be noted: first, the folding has been completely achieved on the vertical plane, and secondly that no interaction between elements occurred during folding.

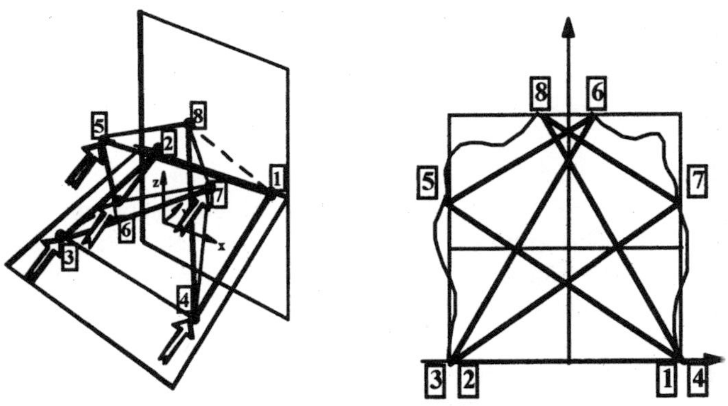

Figure 6.34 *Numerical modelling*

Projections of the trajectories of all the nodes on different planes have been plotted in Figure 6.35. They are very close to those observed with the physical model. At the present time the strut thickness is not taken into account and hence does not influence the final position of nodes.

Node 6 does not move on the vertical plane x = 0 contrary to the geometrical result (because of a simplified assumption). Similar observations can be made for node 8.

Nodes 5 and 7 slide over vertical planes and node 5 approximately along a circular arc. Node 7 reaches the folding plane (y = 0.5) before node 5 (because of the displacement of node 8) and then moves vertically.

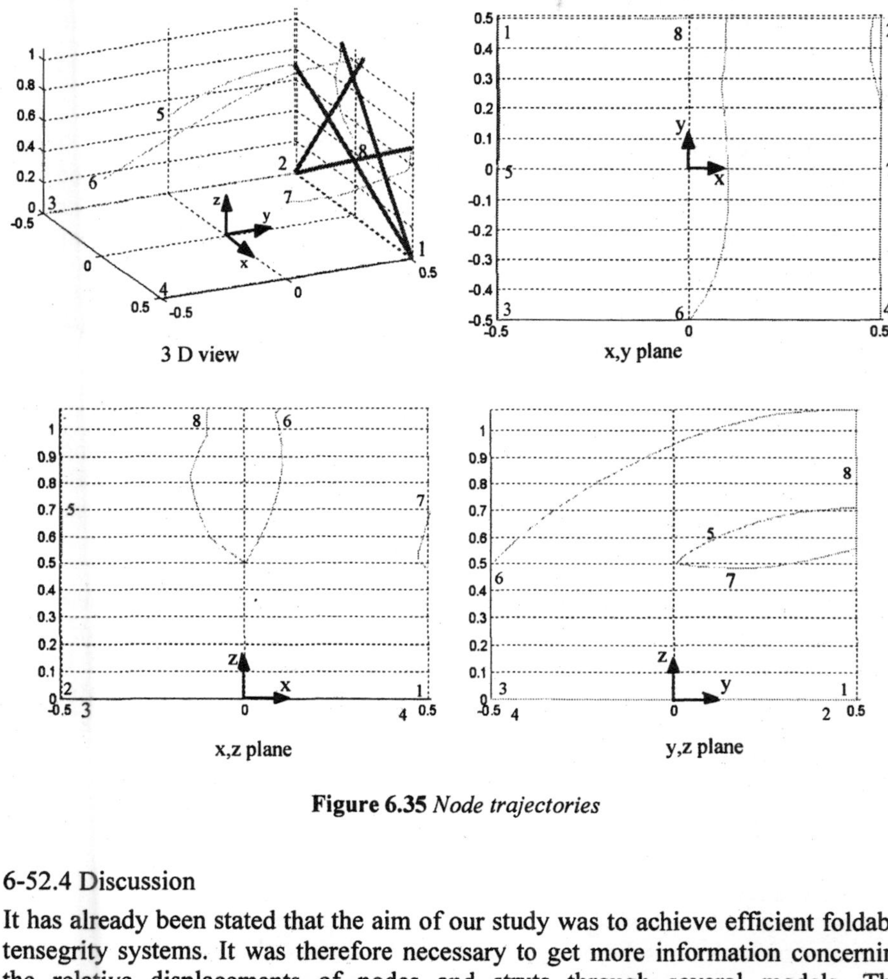

Figure 6.35 *Node trajectories*

6-52.4 Discussion

It has already been stated that the aim of our study was to achieve efficient foldable tensegrity systems. It was therefore necessary to get more information concerning the relative displacements of nodes and struts through several models. This information will provide the data required in order to design the nodes.

Nevertheless, there are some discrepancies between the physical, geometrical and numerical models. This is mainly due to different operating conditions. In fact, in the physical model we cannot neglect the self-weight of the struts, the geometrical imperfections of the nodes and friction in the groove. In the geometrical model the strains in the elements are disregarded and the excessive constraints on the nodes' displacement give trajectories of a "suspect" perfection. Finally, in the numerical and in the geometrical model, the self-weight and the dimensions of the constitutive elements are neglected. The edge-conditions in the last model are close to the physical model.

Consequently, some divergence appears in the results of the above models. The first concerns the trajectory of node 7; as in the trajectory of node 5 in the geometrical model, in the numerical model it is a spatial curve. This fact can be observed in the physical model.

The position of nodes 6 and 8 in the folded geometry is not identical to the numerical model, which is not the case for the geometrical and physical model. It can be explained by the mobility of nodes 6 and 8, in the folded geometry, and of course by the accuracy of the direction of the external actions applied and length of elements.

But excluding the different operating conditions discussed above, there is nevertheless a good correlation between the aforementioned models.

6-5.3 Conclusion

It seems clear that the design of a foldable tensegrity system needs several models to be achieved correctly. The methodology that we adopted uses three of them. Each of the models has its own limits but together they provide the necessary information. The models are complementary.

The physical model shows which cables have to be lengthened, the numerical model allows us to know which cables are strained during the folding process and provides "real" trajectories which can be compared to the "ideal" ones provided by the geometrical model. This methodology has been successfully tested on an assembly of four-strut modules (planar grid) and is currently being tested on other kinds of tensegrity modules and of their assemblies.

6-6 Simulation of the folding process

6-6.1 Introduction

In order to describe a folding process with an appropriate numerical model and its associated numerical resolution, it is necessary to distinguish two main steps, which correspond respectively to the determination of mechanisms and to the description of trajectories implied by the activation of these mechanisms.

6-6.2 Mechanism determination

6-62.1 Basic equations

6-621.1 Cinematic relationships

Static and cinematic equations are established making the classical hypothesis for reticulate structures with struts and cables. Between two different states 1 and 2, the length of member "jk" is dependent on the initial coordinates $\{x_j\},\{x_k\}$ and displacement vectors $\{d_j\}$, $\{d_k\}$.

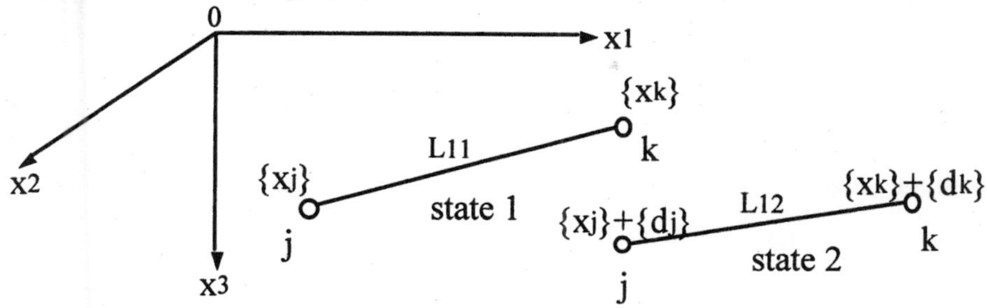

Figure 6.36 *Displacement increments*

According to the notation used in Figure 6.36, it can be written:

E 6.4 $L_{11}= \|x_k\text{-}x_j\|$

E 6.5 $L_{12}= \| x_k - x_j + d_k - d_j \|$

The elastic elongation e_1 can be expressed as:

E 6.6 $e_1 = L_{11} - L_{12}$

or:

E 6.7 $e_1 = [-2(\{x_k\}-\{x_j\})-(\{d_k\}-\{d_j\}),2(\{x_k\}-\{x_j\})+(\{d_k\}-\{d_j\})]^T.\{d_k,d_j\}^T/(L_{11}+L_{12})$

E 6.7 involves not only rigid displacements, but also elastic deformations. Assuming that there are "b" members and "N" degrees of freedom, the cinematic relationship can be expressed in a matrix form as follows:

E 6.8 $\{e\} = ([B]+[\Delta B])\cdot\{d\}$

{e} is the elastic deformation vector; [B] is the compatibility matrix and [ΔB] a increment of [B]; {d} is the displacement vector (boundary conditions being included by deleting the corresponding displacement components). When $\|\{d\}\|$ is very small, the second term can be neglected, then:

E 6.9 $\{e\} = [B]\cdot\{d\}$

For an unstable structure (with finite mechanisms), there is no deformation until the geometrically stable equilibrium state is reached and E 6.9 becomes:

E 6.10 $[B]\cdot\{d\} = \{0\}$

6-621.2 Static equilibrium relationship

Static equilibrium equations can be derived from the principle of virtual work. For a set of external actions {f} and a virtual displacement {δd}, the corresponding elongations {δe} and internal forces {T} must satisfy:

E 6.11 $$\{f\}^t \cdot \{\delta d\} = \{T\}^t \cdot \{\delta e\}$$

Combining E 6.9 and E 6.11 yields:

E 6.12 $$\left(\{f\}^t - \{T\}^t \cdot [B]\right) \cdot \{\delta\} = \{0\}$$

This equation holds for an arbitrary {δd}, so that:

E 6.13 $$[B]^t \cdot \{T\} = \{f\} \quad \text{or} \quad [A] \cdot \{T\} = \{f\}$$

The constitutive law can be expressed in a matrix form as follows:

E 6.14 $$\{e\} = [F] \cdot \{T\}$$

Where [F] is a b-order matrix with:

E 6.15 $$F_{ii} = (L / E.A) \quad i = 1, b$$

6-621.3 Stability criterion

When analysing the structure in a geometrically stable equilibrium-state, we use the compatibility E 6.10 instead of static equilibrium E 6.13. At equilibrium the total potential energy Π of the structure is a local minimum. The necessary and sufficient conditions for equilibrium are:

E 6.16 $$\delta\Pi = 0$$

and

E 6.17 $$\delta^2\Pi$$

Where δ denotes a variation in the displacement space. The equilibrium is instable or stable according to the value of $\delta^2\Pi$. The condition expressed by E 6.16 will be used in the following sections in order to choose a parameter leading to the stable equilibrium-state.

6-62.2 Displacement and self-stress mode matrices

6-622.1 First approach: the four subspaces

E 6.9 and 6.13 describe the cinematic and static relations between two vector spaces associated with a reticulated system space, namely the node space R^n and the member space R^b. External forces and displacements are related to the first one, internal forces and elongations to the second one. If we call "r_A" the rank of the equilibrium matrix [A], a Gaussian elimination procedure on the augmented matrix [A | I], [I] being a diagonal unity matrix, leads to the transformation developed below (Pellegrino, [Ref 6-12]).

E 6.18
$$[A|I]=\left[\tilde{A}\,\middle|\,\tilde{I}\right]$$

which can be put in the form:

E 6.19
$$\begin{bmatrix} \tilde{A}_r & \tilde{I}_r \\ 0 & \tilde{I}_m \end{bmatrix}$$

allowing to transforming the static E 6.13:

E 6.20
$$\begin{bmatrix} \tilde{A}_r \\ 0 \end{bmatrix}\cdot\{T\}=\begin{bmatrix} \tilde{I}_r \\ \tilde{I}_m \end{bmatrix}\cdot\{f\}$$

The external force space {f} can be split into two subspaces: one is r-dimensional and is a fitted external force subspace (forces that can be equilibrated); the second one, of m-dimension, (with m = N-r) corresponds to the forces which activate the mechanisms.

Similar derivations (with [J] being a diagonal unit matrix) can be achieved on the compatibility equation and lead to:

E 6.21
$$\begin{bmatrix} \tilde{B}_r \\ 0 \end{bmatrix}\cdot\{d\}=\begin{bmatrix} \tilde{J}_r \\ \tilde{J}_m \end{bmatrix}\cdot\{e\}$$

The deformation space is composed of two subspaces, a "fitted" and a "non-fitted" subspace. Only the former is compatible with the displacements. If we consider E 6.20 from an energetic point of view, each row of $[I_r]$, and $[I_m]$ can be regarded as a set of displacements and each row of $[J_r]$, and $[J_s]$ as internal forces. In E 6.21, when the components of {d} belong to the mechanism subspace, the corresponding values of {e} are equal to zero and the corresponding rows of $[B_r]$, which are related to the external forces are orthogonal to the displacements. The m mechanism vectors are included in $[I_m]$ from E 6.20 and the corresponding displacements can be computed by:

E 6.22
$$\{d_m\}=\left[\tilde{I}_m\right]^t\cdot\{\alpha\}$$

with $\{\alpha\}$ being composed of arbitrary constants combining elementary mechanisms.

$[D] = \left[\tilde{I}_m\right]^t$ is known as the displacement mode matrix.

A similar analysis leads to the computation of any self-stress vector by:

E 6.23 $$\{T_s\} = \left[\tilde{J}_s\right]^t \cdot \{\beta\}$$

with $\{\beta\}$ being composed of arbitrary constants combining elementary self-stress states.

$[S] = \left[\tilde{J}_s\right]^t$ is known as the self-stress mode matrix.

It should be noted that for a structure, which verifies the compatibility condition, the self-stress subspace is orthogonal to the elastic deformation one. If the structure is in equilibrium, the mechanism displacement space is orthogonal to the external force space.

6-622.2 Second approach: generalised inverse method

Considering the compatibility equation, the matrix [B] is a b x N matrix; generally this is not a square matrix and even if it is square its rank is not always equal to N (= b); we cannot use traditional procedures to solve it. We must introduce the Moore Penrose inverse and, more precisely, the {1}-inverse $[B]^-$. [B] being a b x n dimension matrix, there exists a matrix N x b $[B]^-$called generalised inverse of [B], satisfying:

E 6.24 $$\left([B]^- \cdot [B]\right)^t = [B] \cdot [B]^-$$

E 6.25 $$\left([B] \cdot [B]^-\right)^t = [B]^- \cdot [B]$$

E 6.26 $$[B] \cdot [B]^- \cdot [B] = [B]$$

E 6.27 $$[B]^- \cdot [B] \cdot [B]^- = [B]$$

and:

E 6.28 $$[B]^- = [B]^t \cdot \left([B] \cdot [B]^t\right)^{-1}$$

With this, the general solution of the compatibility equation can be put in the form (Graybill, [Ref 6-13]):

E 6.29 $$\{d\} = [B]^- \cdot \{e\} + \left([I] - [B]^- \cdot [B]\right) \cdot \{\alpha\}$$

with $\{\alpha\}$, arbitrary vector, $[B]^-$, generalised inverse of [B] and [I], identity matrix.

A general displacement {d} can be written:

E 6.30 $$\{d\} = \{d_e\} + \{d_r\}$$

with $\{d_e\}$, elementary displacement-vector, which generates elastic deformations in the members of the structure, and $\{d_r\}$ elementary displacement vector corresponding to rigid body displacements.

In fact it appears that:

E 6.31 $$[D] = [I] - [B]^{-} \cdot [B]$$

which is equivalent to the displacement mode matrix derived from study of the subspaces.

6-6.3 Folding numerical modelling

6-63.1 Compatibility equation solution

When a mechanism basis characterised by [D] has been found, finite mechanisms are activated and this is achieved without any deformation. Then we have to solve:

E 6.32 $$[B]_{i-1} \cdot \{d\}_i = \{e\} = \{0\}$$

which allows us to reach a geometric stable equilibrium state, "GSES" (Liu and Motro, [Ref 6-11]).

The subscript "i" shows that [B] needs to be updated after each elementary nodal displacement. As $[B]_{i-1}$ is a rectangular matrix, the general solution of E 6.32 takes the form:

E 6.33 $$\{d\}_i = \left([I] - [B]_{i-1}^{-} \cdot [B]_{i-1}\right) \cdot \{\alpha\}_i = [D]_{i-1} \cdot \{\alpha\}_i$$

with $[B]_{i-1}^{-}$, generalised inverse matrix of $[B]_{i-1}$, and

E 6.34 $$[D]_{i-1} = [\{\delta_1\} \ldots ; \{\delta_m\}]_{i-1}$$

is the basis matrix of mechanisms in the structure, in the final configuration of step "i-1". Some of these mechanisms may be finite, and others infinitesimal. $\{\alpha\}_i$ is an arbitrary vector; to accelerate the iterative process, it will be determined by energetic considerations.

6-63.2 Energetic considerations

The potential energy variation of the system, under external nodal exterior load vector $\{f\}$, associated to the nodal displacement vector $\{d\}_i$, is:

E 6.35 $$\Delta\Pi_i = -\{d\}_i^t \cdot \{f\} \leq 0.$$

It is judicious to choose $\{\alpha\}_i$ along the gradient of $\Delta\Pi_i$ in order to minimise the potential energy variation. That is:

E 6.36 $$\{\alpha\}_i = \varphi \cdot [D]_{i-1}^t \cdot \{f\}$$

φ is the parameter which defines the increment amplitude for the displacement along the direction of the potential energy gradient. The scalar φ is taken to obtain a final longitudinal deformation of any element smaller than a fixed value ε. This deformation is due to the presence of infinitesimal mechanisms in $[D]_{i-1}$ matrices. With such a φ, the condition "$\|\{d\}_i\|$ small" is still verified. From E 6.33 and E 6.35, it appears that every mechanism of the structure $\{\delta_i\}_{i-1}$ is balanced by coefficient α_i, showing that the potential energy variation of the system, due to the displacement $\alpha_i \cdot \{\delta_i\}_{i-1}$, is minimal.

6-7 Modelling the contact of two struts

6-7.1 Introduction

When one or more contacts occur, a new cinematic restriction imposing the condition that struts in contact cannot penetrate each other is added to the condition that the elements are not allowed to deform and that displacement boundary conditions have to be satisfied, all of which has been previously discussed. The objective of this section is to introduce a way of modelling "Strut-Strut Contacts" during the folding process of tensegrity systems, which has been specifically studied by Le Saux [Ref 6-14].

6-7.2 Motion modelling of two struts in permanent contact

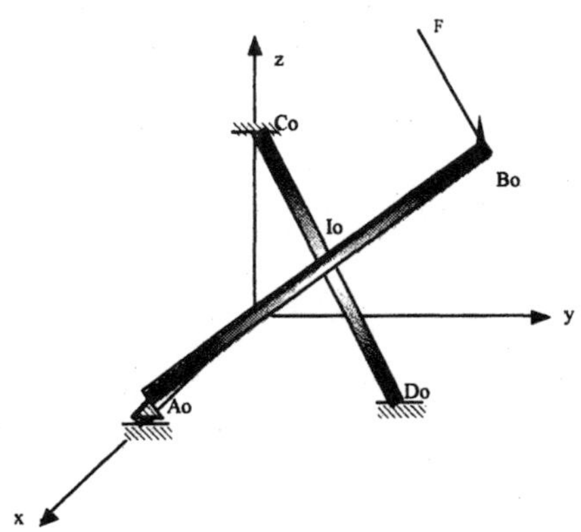

Figure 6.37 *Two struts in contact. Initial configuration (Γo)*

Let us consider two struts AB and CD (numbered 1 and 2 in the following developments) in permanent contact at point I, as in Figure 6.37, but without any boundary conditions and subjected to arbitrary nodal loads. The initial configuration is (Γ_o): (A_O, B_o, C_o, D_o). The numerical model we chose decomposes the motion of the two struts into two parts. The real motion of two struts in permanent contact is in fact a combination of:

- a relative sliding of the struts in contact which implies only independent translations of each strut. This will be called (below), the first part of the real motion;
- a global translation of the struts considered as constituting a single body;
- a relative rotation of struts.

These two movements will be considered together as the second part of real motion. There is no relative sliding in this second part even if a contact point exists.

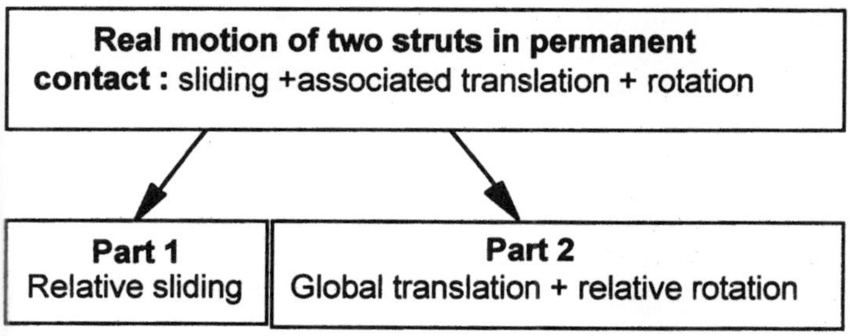

Figure 6.38 *Decomposition of the real motion*

Part 1. It describes the relative sliding of struts in contact, and hence the variation of the contact point between the struts. Sliding is characterised by four independent mechanisms, expressed in terms of nodal displacements.

- a translation of strut CD in A_oB_o direction: $\{\delta_1\}$
- a translation of strut CD in C_oD_o direction: $\{\delta_2\}$
- a translation of strut AB in C_oD_o direction: $\{\delta_3\}$
- a translation of strut AB in A_oB_o direction: $\{\delta_4\}$

The energetic considerations of section 6-63.2 devoted to the modelling of tensegrity folding without contact, lead us to define the four mechanism amplitudes as follows:

E 6.37 $$\alpha_j = \varphi \cdot \{\delta_j\}^t \cdot \{f\} \qquad j = 1,\ldots,4$$

The nodal displacement vector resulting from sliding is:

E 6.38 $$\{d^1\} = \sum_{j=1}^{4} \alpha_j \cdot \{\delta_j\}$$

Thus, we obtain the resulting configuration from Part 1:

$(\Gamma_1) = (A_1, B_1, C_1, D_1, I_1)$.

Part 2. This part models the global translation and the relative rotation motions of struts in contact, without relative sliding. Contact point I_1 does not move on struts AB and CD, but moves with respect to the global reference, by the rigid-body motion of each strut. The latter is defined by three translations and three rotations, relative to the global reference directions (x,y,z). This induces us to define for each strut in contact, a "translation vector" $\{t_j^2\}$ and a "rotation vector" $\{r_j^2\}$, grouped in the $\{{}^*d_j^2\}$ vector as follows:

E 6.39 $$\{{}^*d_j^2\} = \{\{t_j^2\};\{r_j^2\}\}^t$$

The subscript "j" and superscript "2" refer to the "j" strut and part "2" of the motion. "*A" indicates that A is expressed in strut translation and rotation terms.

The permanent character of the contact and the non-sliding is written as:

E 6.40 $$\{d_1^2(I)\} = \{d_2^2(I)\}$$

In other words, the displacement of contact point I, belonging to strut 1, is equal to the displacement of contact point I, belonging to strut 2.

With:

E 6.41 $$\{d_j^2(I)\} = \{t_j^2\} + \{r_j^2\} \wedge \{OI\} \quad j = 1,2$$

Equation E 6.40, projected onto the global reference directions, gives the following system of equations:

E 6.42 $$[{}^*B] \cdot \{{}^*d^2\} = \{0\}$$

with:

E 6.43 $$\{{}^*d^2\} = \left\{\begin{matrix}\{{}^*d_1^2\} \\ \{{}^*d_2^2\}\end{matrix}\right\}$$

$[{}^*B^2]$ is rectangular, so the general solution of system E 6.42 takes the form:

E 6.44 $$\{{}^*d^2\} = \left(I - [{}^*B^2]^- \cdot [{}^*B^2]\right) \cdot \{\alpha^2\} = [{}^*D^2] \cdot \{\alpha^2\}$$

Then the basis of mechanisms for the structure in the (Γ_1) configuration is:

E 6.45 $$[{}^*D^2] = [\{{}_1^*\delta^2\}\{{}_2^*\delta^2\}\ldots,\{{}_m^*\delta^2\}]$$

These are rigid-body mechanisms defined by particular strut rotations and translations. They respect the permanent and non-sliding character of the contact.

With the intention of using the previously developed energetic considerations, to determine the $\{\alpha^2\}$ vector, a linear transformation is used to express $\left[{}^{*}D^2\right]$ in nodal displacement terms: $\left[D^2\right]$

The transformed equation E 6.44 is:

E 6.46 $$\{d^2\} = \left[D^2\right] \cdot \{\alpha^2\}$$

And $\{\alpha^2\}$ is chosen as:

E 6.47 $$\{\alpha^2\} = \varphi \cdot \left[D^2\right]^t \cdot \{f\}$$

So $\{d^2\}$, the displacement vector of all structure nodes, resulting from Part 2 of the motion is:

E 6.48 $$\{d^2\} = \varphi \cdot \left[D^2\right] \cdot \left[D^2\right]^t \cdot \{f\}$$

and the structural configuration, after the nodal displacement

E 6.49 $$\{d\} = \{d^1\} + \{d^2\}$$

is (Γ_2).

Thus an iterative process "$\{d\}_i$" is defined such that the final configuration of the structure $(\Gamma_2)_{i-1}$, at step "i-1" is the initial one $(\Gamma_0)_i$, at step "i".

For each step, the nodal displacement vector $\{d\}_i$ verifies:

E 6.50 $$\left[B\right]_{i-1} \cdot \{d\}_i = \{0\}$$

It means that $\{d\}_i$ does not generate longitudinal deformation of elements in contact:

E 6.51 $$\{e\}_i = \{0\}$$

This is due to the fact that in Parts 1 and 2, only rigid body mechanisms appear.

6-7.3 An example

Let us apply the above model to the problem of a strut AB, sliding on another one CD that is held fixed (Figure 6.37). Node A can move only in the horizontal plane (x,y). At free node B, a load F is applied in the CD direction, it monitors the strut-strut contact. Due to the boundary condition at node A, a cinematic restriction is introduced: equation E 6.52 is this condition and it is added to system E 6.42;

E 6.52 $$\{d_1^1(A)\} \cdot \vec{Z} + \{d_1^2(A)\} \cdot \vec{Z} = 0$$

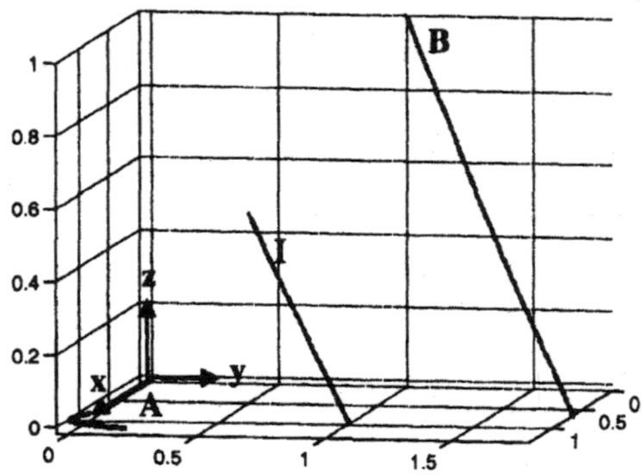

Figure 6.39 *Nodes trajectories 3D view*

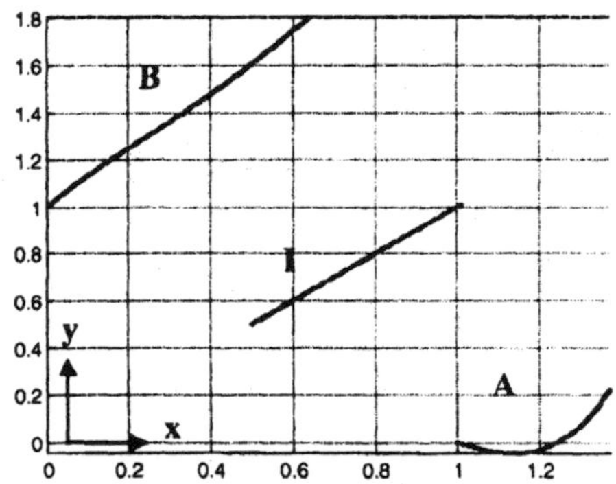

Figure 6.40 *Nodes trajectories xy view*

Examination of Figure 6.39 and Figure 6.40 leads us to conclude that sliding at points A and I has been correctly modelled. A sign change of the Y direction sliding of node A appears during the simulation, which causes the corresponding trajectory to curve. Of course, divergence between the real and numerical trajectories may occur due to the fact that the thickness and self-weight of the elements and also friction at points A and I have been neglected. Nevertheless the results are satisfactory.

6-7.4 Another example

The first stage in the folding of a tensegrity system does not involve any contacts, but the second stage requires the modelling of strut-strut contact. During the second stage, at every step, we must first write the equation E 6.38 for each contact and determine $\{d^1\}$; then we write equation E 6.42 for each contact:

E 6.53 $$[B_{i-1}]\cdot\{d_i\}=\{e\}=\{0\}$$

Boundary condition equations in displacement, such as E 6.52 are then applied. This section presents results given by the numerical simulation of the folding of the four-strut module in Figure 6.41. Cables 1-8 and 3-6 have been lengthened to create mechanisms and then allow folding. As shown in Table 6.1, nodes 3 and 4 are able to slide along the Y direction, and the folding is achieved on the vertical plane containing the three nodes 1, 2 and 8. Moreover, cable 3-4 has been removed to observe the relative motion of nodes 3 and 4, during the simulation.

Table 6.1 *Initial configuration of system with co-ordinates, degrees of freedom (d.o.f) of nodes and external nodal actions*

Nodes	Coordinates (m)			D. o. F.	Forces (N)		
	X	Y	Z		X	Y	Z
1	0.5	0.5	0	none	0	0	0
2	- 0.5	0.5	0	none	0	0	0
3	- 0.5	- 0.5	0	y	0	1	0
4	0.5	- 0.5	0	y	0	1	0
5	- 0.5	0	0.5	x,y,z	0	0	0
6	0	- 0.5	0.5	x,y,z	0	0	0
7	0.5	0	0.5	x,y,z	0	0	0
8	0	0.5	0.5	x,z	0	0	0

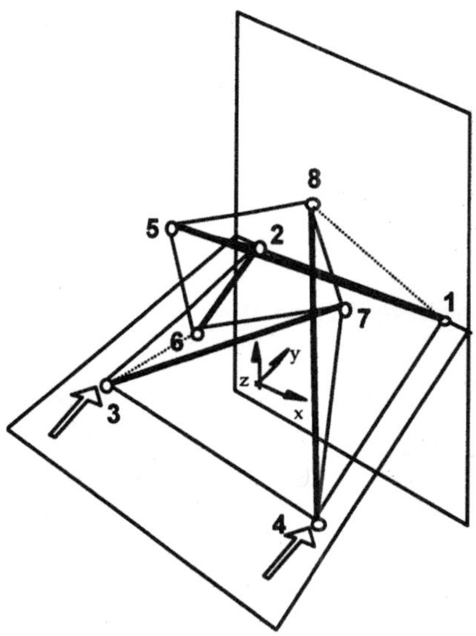

Figure 6.41 *Applied actions on four-strut module*

From the results of the simulation, we can say that at every step of the first stage (without contact), nodes 3 and 4 have the same Y-direction displacement. Then a contact appears between strut 1-5 and 3-7 at point I (node 9 in Figure 6.43), which has a negative x co-ordinate. At every step of the second stage (with permanent contact), the Y-direction displacement of node 4 is higher than the displacement of node 3. This is due to the position of I in the global reference, to the relative positions of struts 1-5 and 3-7, and to the fact that node 3 belongs to a strut in contact, contrary to node 4. Thus, node 4 reaches the folding plane before node 3. The simulation did not show any other contacts.

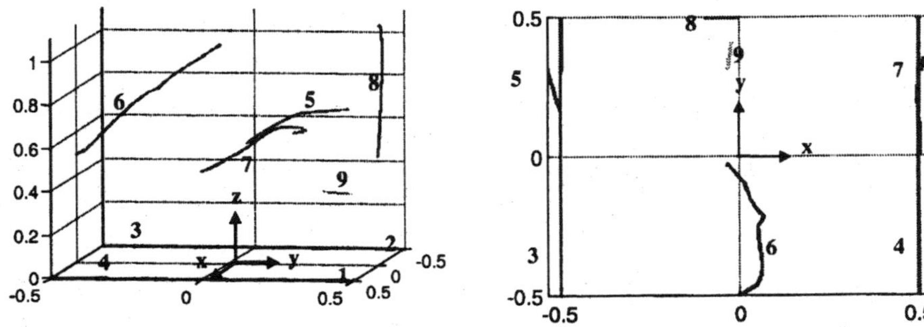

Figure 6.42 *Nodal trajectories during four-strut module folding and folded configuration. First set*

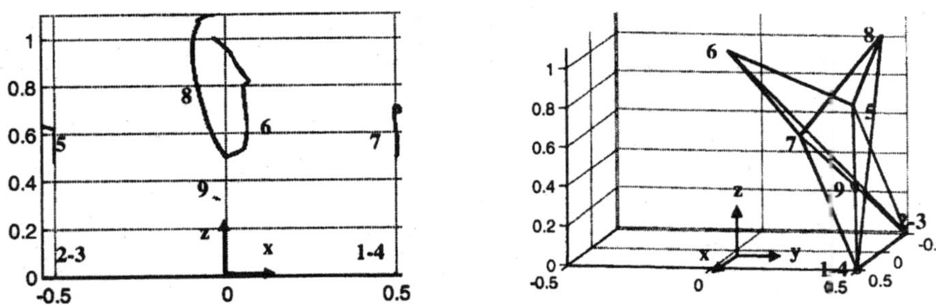

Figure 6.43 *Nodal trajectories during four-strut module folding and folded configuration. Second set*

The trajectories of nodes 5, 6, 7 and 8, each present an irregularity corresponding to the appearance of the contact: they are the highest d.o.f nodes. The contact appears to occur late in the simulation, because only 15% of the total iterations numbers correspond to the second stage.

Results given by the simulation are in conformity with the experimental ones. However, we do not have a real trajectories graphic permitting an accurate comparison with the numerical ones. It is worth bearing in mind that that the thickness and the self-weight of the elements, friction and imperfections in the construction of the system (element fabrication length, play in nodes etc) have all been neglected.

Another folding mode, and one which is more efficient, exists and consists of applying three Y-direction loads at nodes 3, 4 and 6. At the end of the folding, node 6 is indeed in the folding plane (experimental result). However, during the process, two contacts appear between strut "1-5, 2-6" and strut "2-6, 3-7". It means that strut 2-6 participates in two different contacts. W call this type of contact: *related contact*. The existing numerical model can treat several simultaneous contacts only if they are unrelated.

6-8 Conclusion

Needless to say, the work presented here is a first approach in the development of a numerical model for tensegrity systems folding with strut-strut contacts. Nevertheless, this numerical model has produced a number of interesting results such as particular relative motions of some nodes, trajectory irregularity due to contact appearance, and the existence of related contacts.

The challenge is to continue this modelling effort by the realisation of a complete and rigorous numerical model taking into account the following: self-weight, thickness of elements, friction, related contacts and breaking of contacts. It should provide useful information for the technological design of nodes and the development of efficient foldable tensegrity systems. For this goal it is also helpful to build some physical models.

Nevertheless, the future field of tensegrity systems is partly an appropriate combination of active control and folding process.

References

Ref 6-1 Motro R., Bouderbala M., "Mobile Tensegrity Systems", in Proceedings of MARAS 96 "Mobile and Rapidly Assembled Structures II", 1996, pp. 103–113.

Ref 6-2 Vassart N., Laporte R., Motro R., "Determination of mechanisms's order for cinematically and statically indeterminate systems", *International Journal of Solids and Structures*, Vol. 37, 2000, pp. 3807–3839.

Ref 6-3 Calatrava S., Sobre la plegabilidad de entramados, In Arquitectura Transformable, Publicacion de la Escola Tecnica Superior de Arquitectura de Sevilla, 1993, pp. 33–93.

Ref 6-4 Kwan A.S.K, You Z., Pellegrino S., "Active and passive cable elements in deployable/retractable masts", in *International Journal of Space Structures*, Vol. 8 N° 1/2, 1993, pp. 29–40.

Ref 6-5 Kwan A.S.K, Pellegrino S., "Design and performance of the octahedral deployable space antenna (Odesa)", in *International Journal of Space Structures*, Vol. 9 N° 3, 1993, pp. 163–173.

Ref 6-6 Hanaor A., "Double-layer tensegrity grids as deployable structures", in *International Journal of Space Structures*, Vol. 8, N° 1&2., 1993, pp. 135–145.

Ref 6-7 Furuya H., "Concept of deployable tensegrity structures in space application", in *International Journal of Space Structures*, Vol. 7 N° 2, 1992, pp. 143–151.

Ref 6-8 Motro R., Formes et Forces dans les Systèmes Constructifs. Cas des Systèmes Réticulés Spatiaux Autocontraints, Thèse d'Etat, Université Montpellier II, June 1983.

Ref 6-9 Bouderbala M., Systèmes spatiaux pliables/dépliables: cas des systèmes de tensegrité, Thèse de doctorat, Université de Montpellier II, 1998.

Ref 6-10 Hangai Y., "Analytical procedure for stabilising paths and stability of cinematically indeterminate frameworks", in Proceedings of International Association for Spatial and Shell Structures Symposium, Copenhagen, 1991.

Ref 6-11 Liu Y., Motro R., "Shape analysis and internal forces in unstable structures", Proceedings of the IASS International Symposium 1995 "Spatial Structures: Heritage, Present and Future", ed. by G.C. Giuliani, SGE pub., Padovia, Vol. 2, 1995, pp. 819–826.

Ref 6-12 Pellegrino S., Calladine, "Matrix analysis of statically and cinematically indeterminate frameworks", in *International Journal of Solids and Structures*, Vol. 22, 1986, pp. 409–428.

Ref 6-13 Graybill, Introduction to matrices with applications in statistics. Wrels Worth Publishing Company, Inc., Belmont, California, 1969.

Ref 6-14 Le Saux C., Modélisation numérique du pliage des systèmes de tenségrité avec prise en compte des contacts, Mémoire de DEA, Université de Montpellier II, 1998.

7

Tensegrity: Latest and Future Developments

7-1. Introduction

It is now over half a century since the earliest work relating to tensegrity was carried out, which leads one to ask what its future might be. It appears that a number of current works demonstrate the feasibility of these types of structure and their application to the world of architecture has become a reality. For instance we have ourselves developed new designs of tensegrity grids, which are described with other projects and are in the process of being built.

Tensegrity systems are characterised by initial states of self-stress. We described in Chapter 6 the foldability of tensegrities. These two features give access to the development of smart tensegrity structures: it is possible to monitor the level and distribution of self-stress in the deployed state.

Finally, tensegrity can be considered as a specific state independently of its field of application. For it constitutes a model with its own properties and has thus become a structural principle that can be used in fields other than architecture.

7-2. New tensegrity grids

7-2.1. Introduction

Since early 1998, a three year research & development project, called "Tensarch", was launched by our own team at the Mechanics and Civil Engineering Laboratory in Montpellier, France. It was sponsored by a company called Ferrari – the leading fabric membrane producer in France. We had to study the real possibilities of introducing tensegrity in architecture, as a genuine building system, and also to produce a basic industrial design of a tensegrity grid, of about 100 square meters, with a feasibility study. To sum up the key question was: *'What kind of tensegrity and what of its applications in architecture?'*

This project constituted the main part of the thesis of Vinicius Raducanu [Ref 7-1]. Several illustrations in this section are reprinted from his work.

7-2.2. Design considerations

Generally previous grids were designed by agglomeration of self-stressed cells put together. This was true for one of our projects (Figure 7.1).

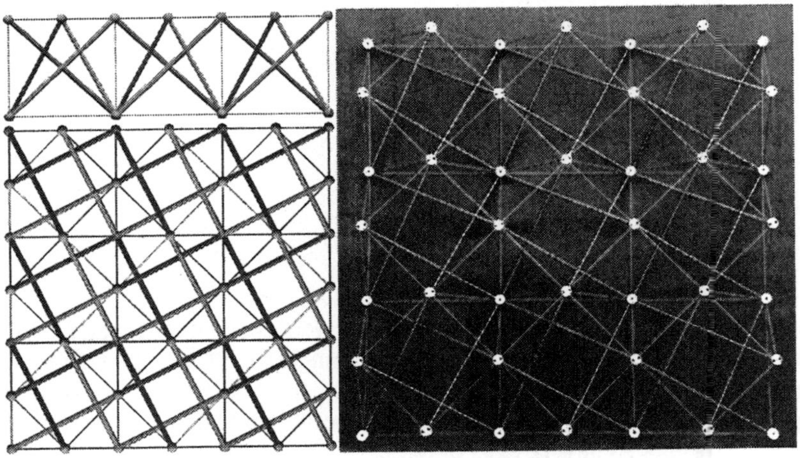

Figure 7.1 *Double layer tensegrity grid (R. Motro, 1990)*

The decomposition of this type of grid into separate compressed components has been described in Chapter 2 "History and Definitions".

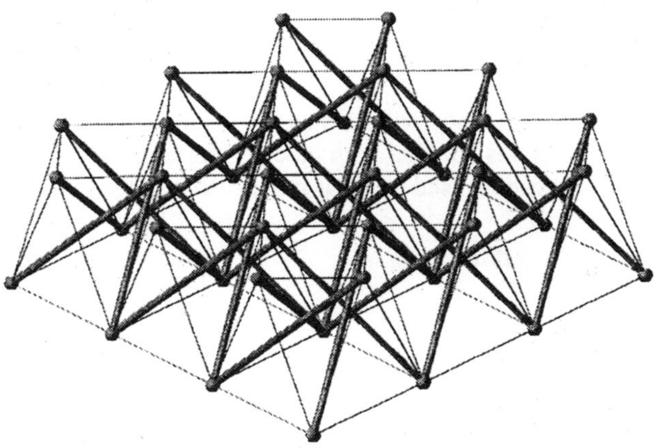

Figure 7.2 *Double layer tensegrity grid: axonometric view*

The previous grid design methods (see [Ref 7-2][Ref 7-3][Ref 7-5]) by proliferation of a self-stable tensegrity unit, kept the original infinitesimal mechanism of this and generated a permanent lack of stiffness. These twisted prisms (general shape of a tensegrity *simplex*) are tiled upon intricate patterns with strut/middle-of-cable jointing solutions. In true symmetrical grid solutions, interaction balance does not require self-sufficient subsystems. Our first aim was to find this kind of tensegrity grids.

It seemed obvious that another design approach had to be employed. Analogy, geometry, topology, static, none of these points of view is able to independently produce genuine new layouts – whereas a synthesis does.

7-2.3. Tensarch project

The complete description of the Tensarch project is given in the PhD dissertation written by Raducanu. Some key ideas are nevertheless set out in this section. Our intention was to design grids by another way – rather than via the agglomeration of self-stressed cells.

In order to avoid the problems that have until recently prevented the employment of tensegrity in architecture, we planned a specific checklist for new flat tensegrity grid layouts:

- A single pattern for all linear components – i. e. each element's projection on the grid plane will be a side of one and the same regular tiling of this plane (simplifies the structure).
- Minimal strut density over a fixed covered area (weight gain).
- No mechanisms (infinitesimal nor finite) and stiffness which match building standards.
- Simplified pre-stressing process, with about one tenth of active elements among the total number.

7-23.1. Analogy

Many scientific papers have compared tensegrity systems with a football – the rubber membrane acts as the continuous tensioned cable net, and the inside air as the struts do, the whole achieving a state of self-stress. In the case of a tensegrity grid, we have put forward a new analogy, that of *the spring mattress*. It is composed of a regular array of isolated spiral springs, tied up by continuous knotted strings and compressed by supplemental transversal strings (the *active* elements in pre-stress), that go right through the mattress, binding the upper and lower faces. A tensegrity grid may be considered as a kind of spring mattress, regarding the similar external behaviour and internal layout ("*islands of compression in a ocean of tension*").

Each of these systems is a whole made by an outer flexible surface, tensioned by including tandems of tensioned and compressed devices (so it does not necessary comprise autonomous small parts – the basic tensegrity systems). The self-stress may be introduced by acting only on some internal components. The external flexible "skin" of the tensegrity grid can be made by reticulate nets but also by continuous sheets.

7-23.2. Constitutive approach

The spring mattress analogy led us to an intuitive but overall vision of a tensegrity grid outlook (see Figure 7.3), bringing into question the nature of a tensegrity grid, beneath the simple way of disposing struts and cables in a self-stress state.

1st. The tensegrity grid is bordered by *layers* (see 2nd pt) and contains inside *"expanders"* (see 3rd pt), the whole making a statically stable system, thanks to mutually balanced efforts.
2nd. Each of those layers is a *surface net* (bi-dimensional entity), totally *flexible* (formed by linear, rigid-only-in-tension components) and *regular* (the elements are spread out over a projection of a regular tiling of the plane). The junctions of the linear elements form *nodes*, the only contacts between layer and extenders.
3rd. Identical subsets of struts and cables, named *expanders*, are distributed inside the grid, between the layers, joining up a node cluster of one layer to another node cluster of the other. In each layer, the expander spreads the nodes of a cluster away from one another. The same device keeps the two groups of nodes at a stable relative distance.
4th. Any node belongs to at least one expander.

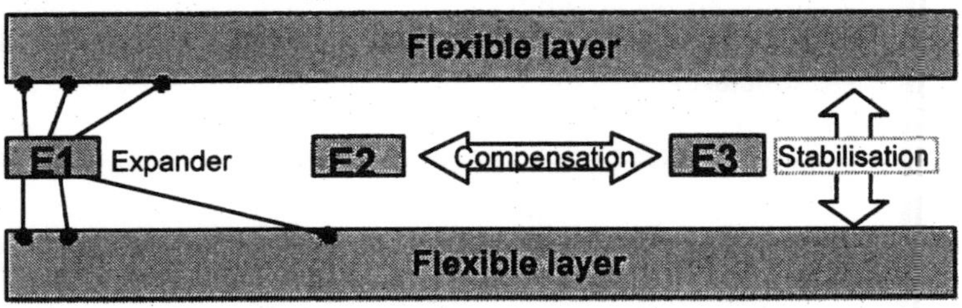

Figure 7.3 *General constitution of a tensegrity grid*

We gave a shape to those concepts, layers and expanders, using geometry and topology.

7-23.3. Geometric approach

The study of regular, plane or space tiling is the first geometric tool which dictates global layout, as was the case for reticulate space frames, generating linear elements with the same plane incidences and the same lengths. With the same stress in each linear element, the entire frame should be in static balance, thanks to its geometric regularity. Consequently, we made a powerful hypothesis: assembling mutually balanced frames of tension and frames of compression (basic ingredients of a self-stress state). *They are geometrically superposed on the same regular tiling.* This should lead to a system of describing nodes stabilised by this self-stress state – and the whole making a tensegrity system. Our hypothesis is of an *infinite* frame and, moreover, we will have to solve border problems as soon as we 'cut' the frame on a finite number of modules. Firstly, by disposing on the same tiling different tension or compression frames of elements, we needed another tool to avoid their intersection: *topology*.

7-23.4. Topological approach

R. B. Fuller's "*islands of compression*" may be an articulate frame of struts, avoiding any tension or bending effort. We will retrieve the fundamental tensegrity principle by assembling many topologically isolated islands of this kind inside a continuous cable net. Struts may be connected as a weaving of isolate, open paths of struts (Figure 7.4); as a single network of adjacent, closed circuits of struts; as multiple, intertwined or superposed networks of the same kind. Combined with geometric generations, about a dozen stable tensegrity grids were submitted for checking by computer.

Figure 7.4 *Weaving of compressed components*

7-23.5. Models and computer analysis

A computing procedure (including home software "*Tenségrité 2000*", see [Ref 7-6]), indicated self-stress states and internal mechanisms, both with complete 3D modelling and imagery. The final step of this constitution-finding process was to carry out five physical models of about 0.2 × 2 × 2m each. We evaluated the level of difficulty and "policies" of the assembling and pre-stressing process for different grid layouts, before choosing the large industrial prototype.

7-23.6. A new grid configuration

Based on the same mechanic principle (the *V expander*) but applied to different geometry, the new grids are named according to the number of distinct directions of strut proliferation onto the grid's plane.

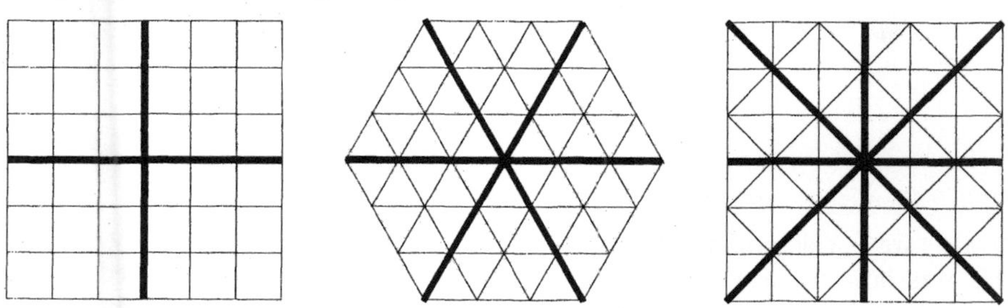

Figure 7.5 *Directions of strut proliferation for bi-, tri- and quadri-directional tensegrity grids*

7-236.1 The "V expander"

One basic 'device' that generates self-stress states for many new tensegrity grids, including the bi-directional, is the V expander, constituted by:

1. Two struts, each one converging to a pin jointed node, placed respectively on either grid's flexible layer. The expander's axis, joining this couple of nodes, is normal to the layer's surface. 2, 3 or 4 struts, uniformly distributed around the expander's axis can constitute any expander.
2. An active cable materialising the expander's axis (if one considers a horizontal grid, this cable will be vertical). By reducing its length, it is tensioned and meanwhile it introduces a self-stress state among some elements of the grid, in its neighbourhood.
3. Many layers' cables, joining respectively the converging node of one pencil of struts to the nodes on the free ends of the other pencil's struts. Thus, those cables belong to the layer and the expander simultaneously. Their number is equal to the number of the struts forming the expander.

The bi-directional grid presented here is composed of a 2V expander (2 + 2 struts), shown in Figure 7.6.

Different expander principles were also imagined; they are described in [Ref 7-1].

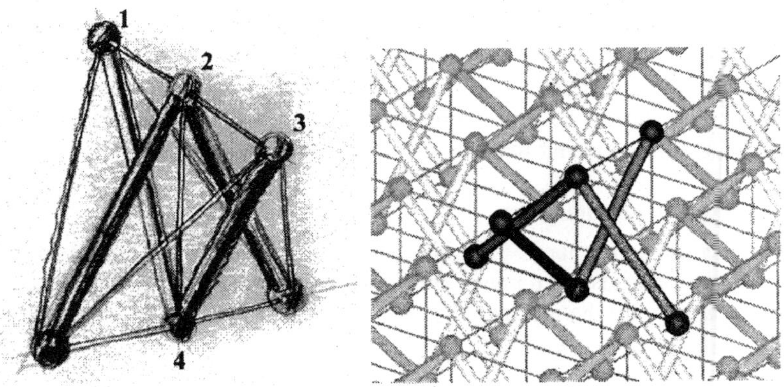

Figure 7.6 *2V expander*

7-236.2 Elementary stitches

Another way of describing the grids is their decomposition in elementary stitches. This was how we named regular prismatic subsets, formed by cable edges and containing struts on some diagonals. These stitches have no structural stiffness in an isolate situation because of, on the one hand, unbalanced lateral thrusts provided by diagonal struts and, on the other hand, finite mechanisms. Nevertheless, assembling these stitches by symmetric duplication along each of the vertical faces results in statically stable systems by providing appropriate border solutions (e.g. specific edge cables) which balances the lateral thrusts of the struts.

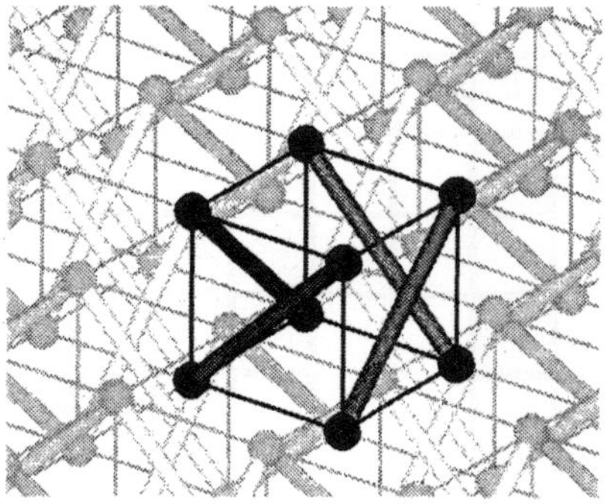

Figure 7.7 *Cubic elementary stitch*

Figure 7.7 shows a regular cubic stitch (that generates the bi-directional grid), but we may also have irregular proportions and/or bevelled forms (for example, they may become truncated pyramids). This flexibility allows the creation of grids adapted to various plan implementations and/or curved general shapes. For the sake of clarity however, we will retain here only the regular shape, which generates only flat grids on regular plane tiling. It is interesting to point out that the elementary stitch is the well-known, four-strut simple tensegrity system "straightened up".

<u>7-236.3 The bi-directional tensegrity grid</u>

This tensegrity grid layout is of an astonishing simplicity (see the plane view in Figure 7.8). We can describe this grid either by its 2V expander or by its cubic elementary stitch (see Figures 7.6 and 7.7), but also by *frames*. It is made up of repeating frames of struts and cables, disposed alternatively upside down and upon two directions. One of these frames is apart on the following Figure 7.9. The structure depicted here is also formed by 8 × 8 cubic stitches. So, all the layer cables and vertical cables have the same final stretched length. On the periphery, each extreme end of one strut (i.e. frame) is bound by two oblique bracing cables to the neighbouring strut ends, which are actually placed on the other side of the grid. Thus, all of those edge cables are forming a continuous zigzag pattern along the border of the grid, visible in the upper middle zone of Figure 7.10. This cable acts as a shoelace for the grid, by "bringing nearer" the borders of the two cable layers.

This plane view may reveal the finite mechanism of each frame: a relative displacement of the frame along its longitudinal axis. Specific support cases or supplemental diagonal border cables (not shown here) preclude the activation of this mechanism. Diagonal bracing solutions are similar to those applied for stabilising orthogonal framing structural systems.

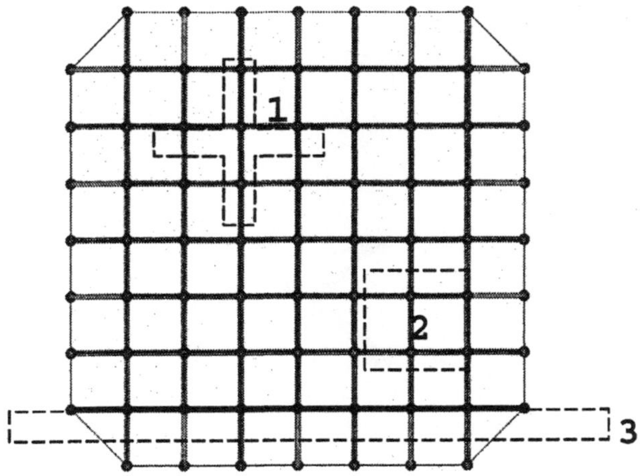

Figure 7.8 *Plan view of the bi-directional grid; surrounded constitutive parts: 1) a 2V expander, 2) an elementary cubic stitch, 3) a frame*

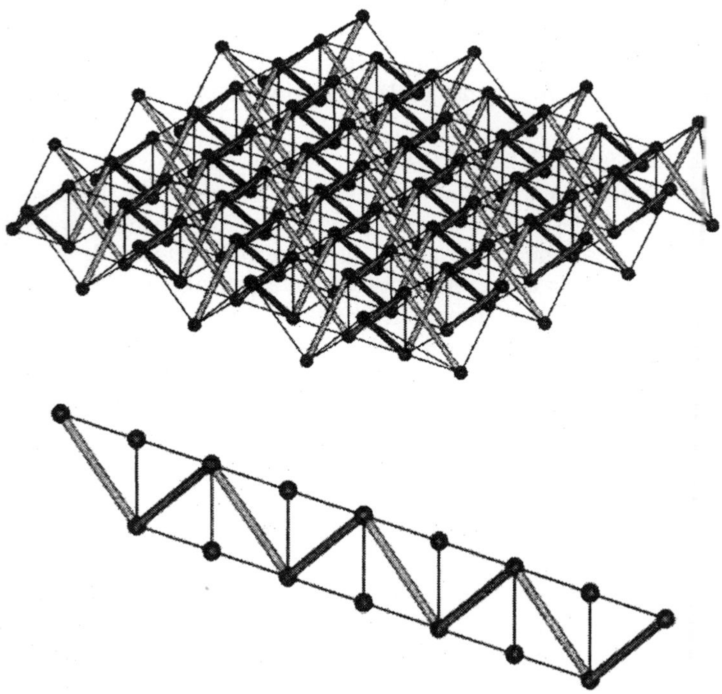

Figure 7.9 *From top to bottom: axonometric view of the bi-directional grid; one of the repetitive frames*

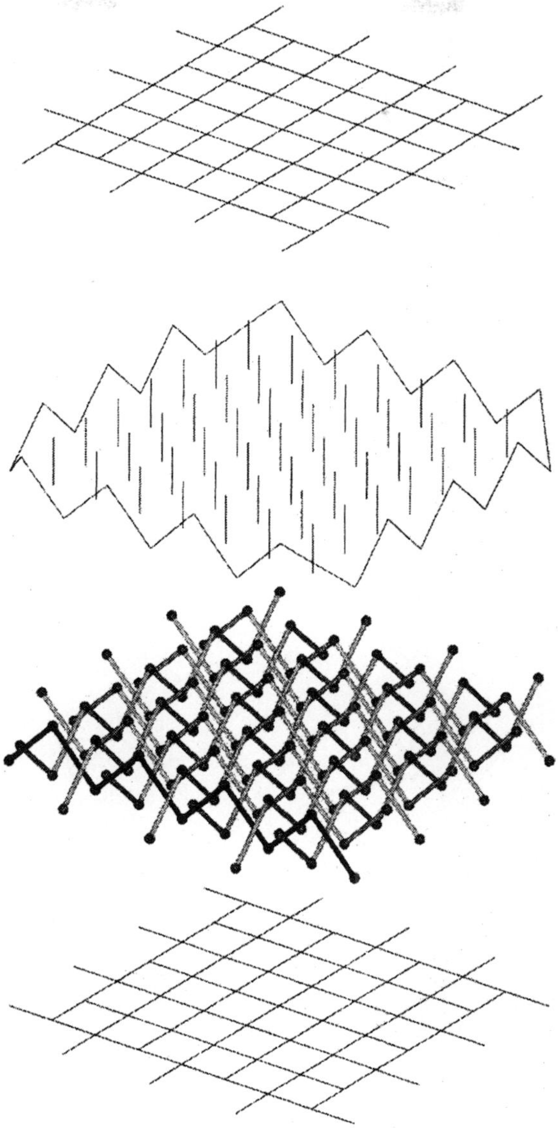

Figure 7.10 *Split axonometric view of the bi-directional grid; from top to bottom: the upper layer of cables, the bracing cables, the taffeta-woven struts, the lower layer of cables*

It should be noted that, topologically speaking, each frame contains a single continuous path of struts from one end to another. This subset of struts constitutes one of the isolate compressed components of the grid. The reason is that the way of crossing between any two perpendicular frames is done by creating a 2V expander at the intersection axis. Respecting this kind of crossover each time, the general pattern of strut subsets becomes a taffeta-style weaving.

Uniformly reducing the length of each vertical bracing cable performs the self-stress implementation. This reduction activates the closest expander and stiffens a local grid area. In this way, one acts on the minimal number of elements.

7-236.4 Prototype

Among the new tensegrity grids, the bi-directional layout is certainly the simplest, both for understanding and prototyping. Therefore, a first industrial grid of this kind was created by the LMGC[1] for the Tensarch Project during the winter of 2000–2001. Covering a $82m^2$ area and weighing 900kg, this steel structure was designed according to the Eurocode3 building standard for a $160daN/m^2$ external downward load and peripheral supports.

Figure 7.11 *Assembled prototype of a bi-directional grid at Nîmes, France, in December 2000*

[1] LMGC: Laboratoire de Mécanique et Génie Civil (Mechanics and Civil Engineering Laboratory).

Figure 7.12 *Assembly at node*

Specific nodes were designed for this project. Two cable layers go through the nodes, which also receive two struts and one vertical tensioned component (Figure 7.12). A detail of strut link with the node is also depicted in the same figure.

All details were first tested on a so-called "mini grid" (Figure 7.13).

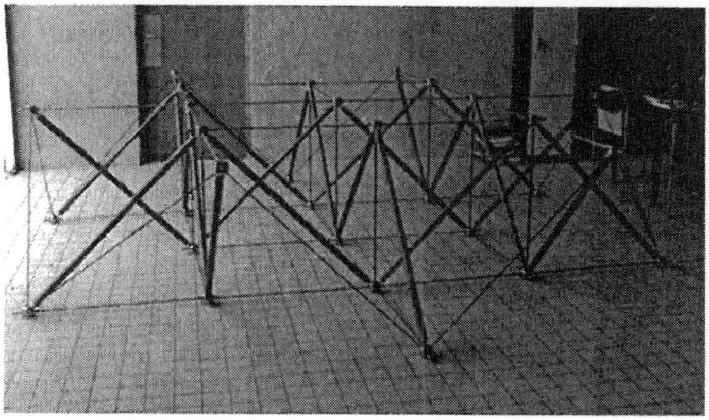

Figure 7.13 *"Mini grid"*

This project assessed the feasibility of simple tensegrity systems; implementation of self-stress was controlled on the mini grid with a precision under 3%. Other tests are currently being carried out on the whole grid, the rigidity of which is surprising (Figure 7.14).

Figure 7.14 *Tensegrity grid in lifted position*

7-2.4. New grids

Other grids based on the same concepts of symmetrical expanders and plane tiling were also imagined. A patent protects those set out here since 2001 [Ref 7-7]. The three main types can be compared with classical space structures according to the orientation of strut paths: two, three and four way grids are represented below. For each of them we have given a view of expander, a plane view and an axonometric view.

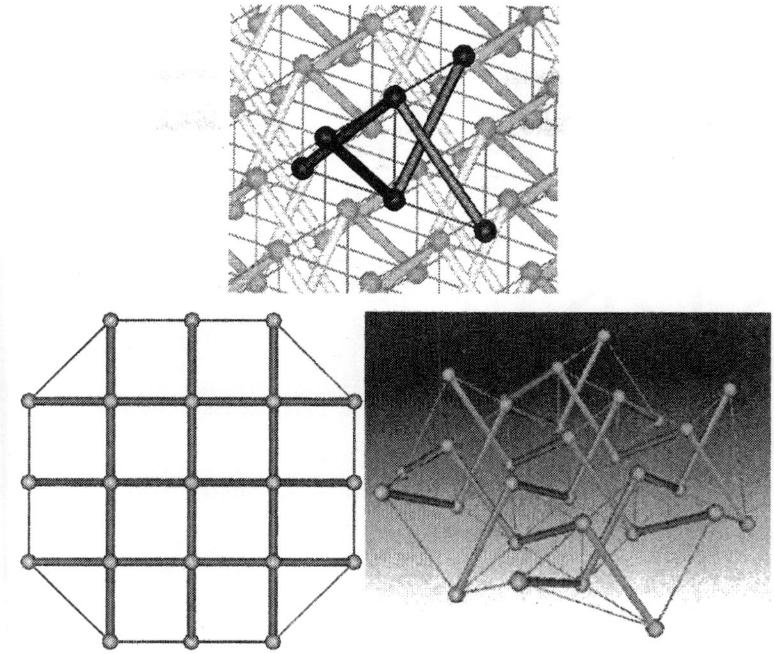

Figure 7.15 *Two ways grid*

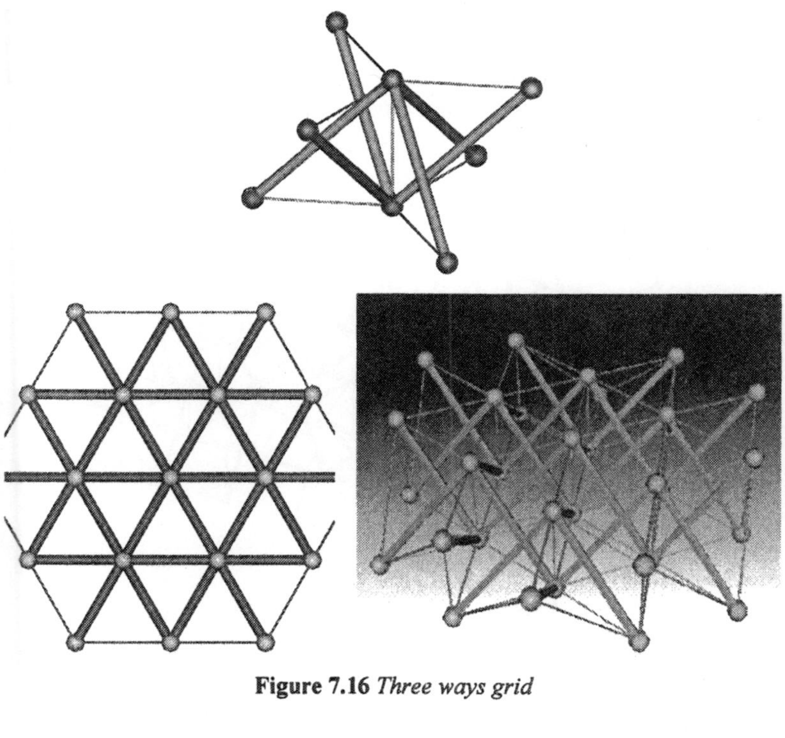

Figure 7.16 *Three ways grid*

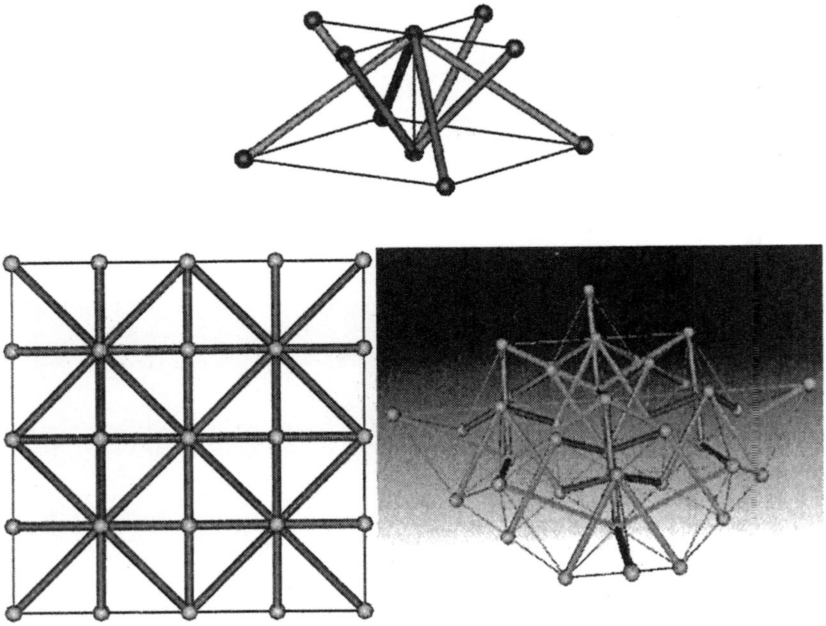

Figure 7.17 *Four ways grid*

It is possible for the three cases to check the definition of tensegrity that we submitted at the beginning of this book. For example, the following decomposition can be made for the four way grid. The whole system in Figure 7.18 contains a continuum of cables and three compressed components, each being achieved with a set of struts, and these three sets do not touch each other.

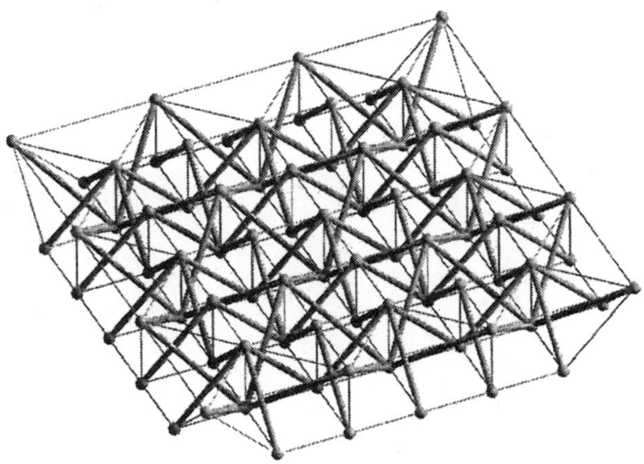

Figure 7.18 *Complete four ways grid*

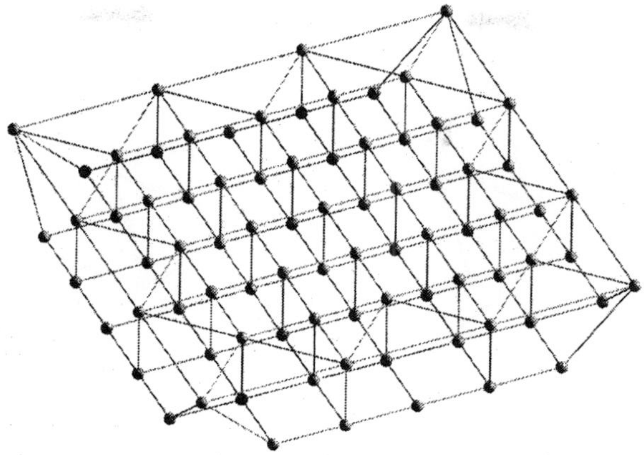

Figure 7.19 *Continuum of tensioned components*

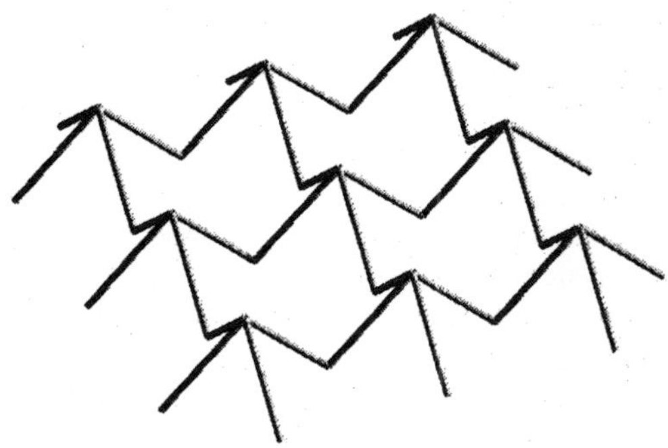

Figure 7.20 *First compressed component*

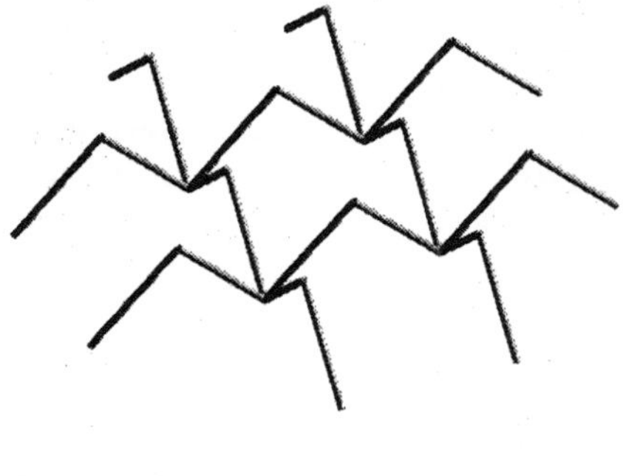

Figure 7.21 *Second compressed component*

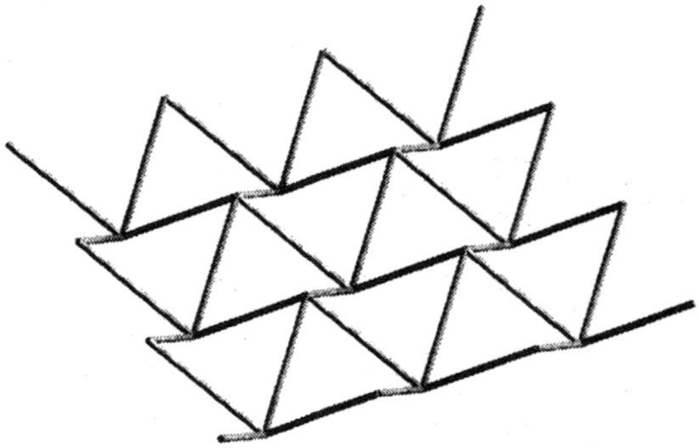

Figure 7.22 *Third compressed component*

7-2.5. Conclusion and perspectives

An important step was reached by conceiving these simple tensegrity grids, together with proper self-stress calculus and implementation. At this stage, a wide range of developments has opened up: conceptual, morphological, technological and, one hopes, architectural achievements too.

The different understandings of tensegrity systems that occurred in the design process, as 'discrete pneumatic structures', enabled us to propose a general definition of tensegrity systems [Ref 7-4]. The by-layer composition of the grids enabled the concept of modular design for optimised nodes. The regular array of elementary stitches allows fluid passing integration in multifunctional wall solutions. Finally, one may consider the tensegrity grids as reliable, solid, adaptive and *quite* rigid discrete pneumatic structures.

Several morphological developments are in the process of being studied:

- **rotated** 'cutting pattern' (e.g. a bi-directional grid with struts oriented at 45° in a rectangular plan of the grid) in order to achieve more structural efficiency.
- possibility of introducing **holes** with the same bracing cables as the exterior borders of the grid, enabling access or natural lightning.
- introducing simple or double, local or general **curvature**, in grids with constant or variable thickness: a lenticular shape for a square grid, or a waveform shaped grid with constant thickness. For a general curvature, the actual node positions may be found by simply projecting the grid's tiling upon the desired layer surface, and then summarily check the static of each type of node (i.e. the correct incidence of struts inside the solid angle formed by the cables).

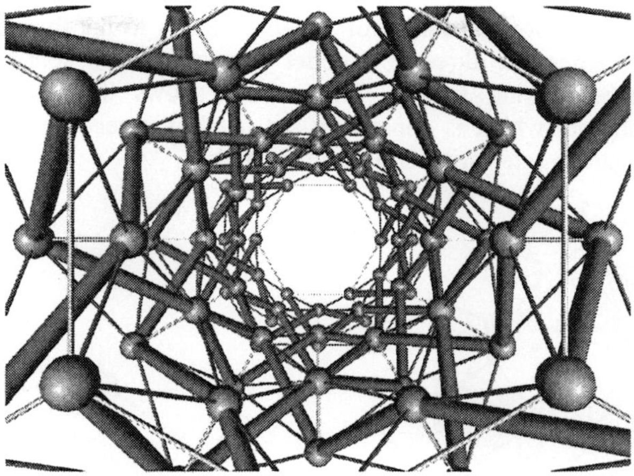

Figure 7.23 *A tensegrity tube, based on a 'rolled' bi-directional grid connected on itself*

- **irregular tiling** which permits complex area coverings, as trapezoidal or diamond shaped. This is now an obvious development, but not so long ago, the strict equilibrium of tensegrity basic systems did not allow any 'deformed' shapes for covering 'any' polygon by assembling them in grids.
- **structural connection** by interpenetrating two or more similar grids, in order to create rigid border links between them (for example with continuous strut paths from one side/grid to the other).

We also have to answer other fundamental questions: what building parts will be able to be carried out in accordance with tensegrity principles, at a given scale – centimetre, metre or decametre? What material associations and jointing solutions must be employed for each of these scales? Giving appropriate answers to these new questions will offer a wide field of application to tensegrity.

7-3. Other projects

Until recently very few projects with tensegrity systems have been developed. Some have been inspired by concepts such as cable domes, but we soon pointed out why they could not be considered as tensegrity systems.

Nevertheless, in recent years many publications[2] have appeared on the subject of tensegrity systems: many are theoretical and contribute to the knowledge in this field. Since it is always difficult to know all of the realised projects in any field, I will only point out some of them since they appear to be indicative of current trends.

7-3.1. Dynamic study of double layer grids

One project has been developed by Kono *et al* [Ref 7-8]. It has already been described in a previous chapter. Its constitutive cell is given in Figure 7.24. It is a

[2] See the Bibliography.

modified simplex the lower triangle of which is linked to the three nodes by three supplementary cables: the three new nodes are common to contiguous cells and create an interpenetrating system. Figure 7.25 shows the realisation with an associate membrane, which is apparently not used as a resistant component. The authors carried out an interesting and dynamic study.

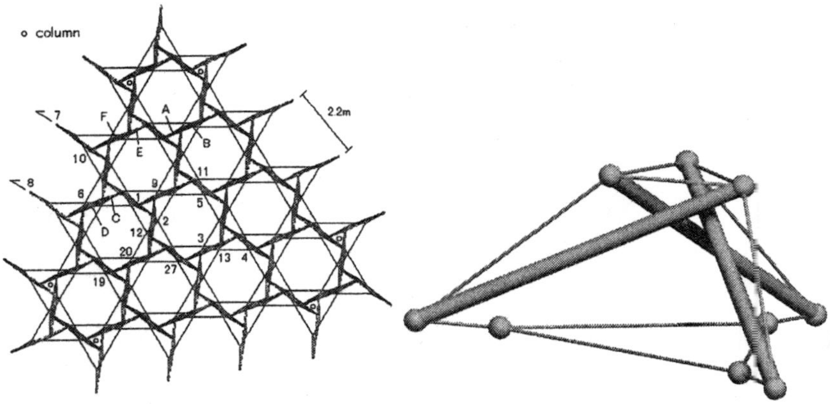

Figure 7.24 *Plane view and constitutive cell*

Figure 7.25 *Kono's project*

7-3.2. Passera and Pedretti projects

Another work constitutes a major step in the field of tensegrity systems. It has been carried out by Passera and Pedretti who worked for international Swiss Expo '02.

They built many models and designed a specific node. A prototype has been erected in Lugano, but until now it has not been loaded. Another prototype has also been built in Lausanne at the Polytechnic School with fewer cells; it is used for active control studies by I. Smith and E. Fest [Ref 7-9] and will be presented in the next section. The entire work completed by these engineers constitutes, to the best of my knowledge, one of the first "intrusions" of tensegrity in engineering. Effective projects will certainly follow these interesting early studies. Passera and Pedretti are currently constructing two projects based on their experience gained in the field of tensegrity.

The first project concerns directly a so-called "Arteplage". Mauro Pedretti decided to call his structure S3 that corresponds to Strut-String Structures, since he was not convinced at this time that this structure was a tensegrity system.

Figure 7.26 *EXPO '02 Arteplage Yverdon-Les-Bains – Ch*

The second project has been designed with Diller & Scofidio (Architecture New York) with Passera Pedretti.

The form of the roof of the Velodrome is an ellipse. The major and minor axes are 90.8m and 67.1m, respectively. The upper part of the roof is a pneumatic structure made of two layers of fabric membrane. The pneumatic structure is supported on a ring of steel Strut-String Structure, which is composed of "**Bi-Pyramidal Tensegrity Modules**". The covered area of the roof is 4780m2. The unit weight of the roof is 34kg/m^2.

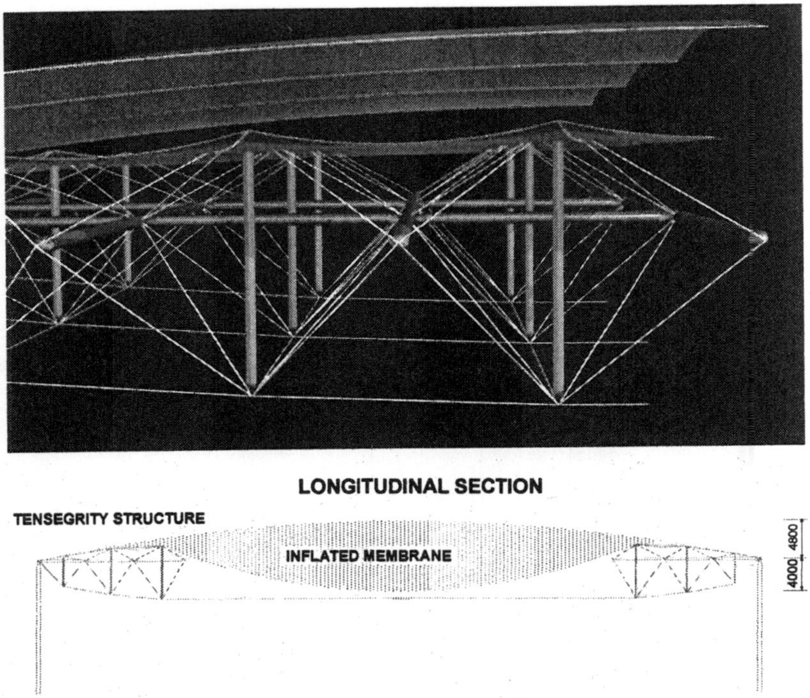

Figure 7.27 *Centre Mondial Du Cyclisme A Aigle – Ch*

These two projects are currently under completion.

7-3.3. Tor Vergatà arch

We established a close scientific collaboration with some Italian scientists who are also working on tensegrity. We had a first colloquium in Rome (May 2001). This scientific program is supported by a research partnership between several laboratories both in Italy and in France ("Groupement de Recherche Lagrange"). The Dean of Tor Vergatà University had decided to build a tensegrity arch on the campus and asked Paulo Podio Guidugli to design the project. The first idea for this arch with a clear span of about 50 m is to assemble elementary six-strut cells like the one called "expanded octahedron". This composition rule has already been experienced on a smaller scale during a workshop on tensegrity that I held at the special school of architecture in Paris in 1999 (Figures 7.28 and 7.29). We hope to carry out this project in the next few years and use it as a real physical model to estimate the effective actions of wind.

Figure 7.28 *Tensegrity arch: side view*

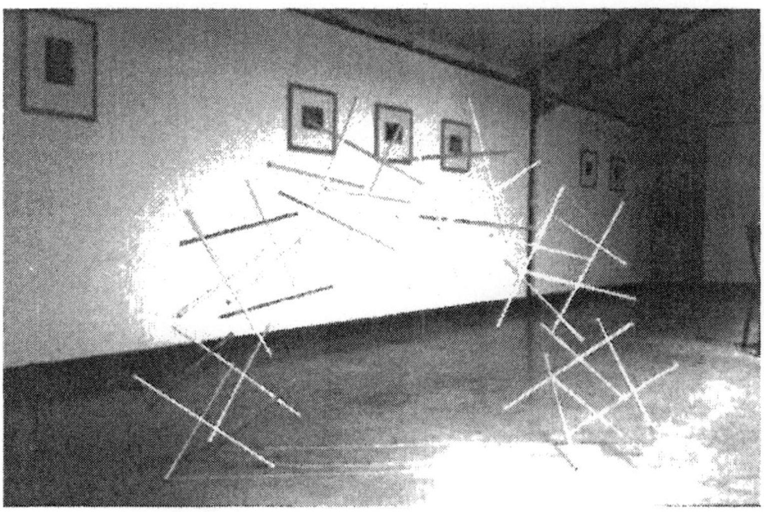

Figure 7.29 *Tensegrity arch: front view*

7-3.4. Smart structures

7-34.1. Basic ideas

E. Freyssinet explicitly introduced pre-stress at the beginning of the twentieth century (his patent goes right back to 1928). This famous engineer applied pre-stress mainly for concrete structures. Some years later, engineers began to think about pre-stressing other materials. Bernard Laffaille was one of the pioneers in the field, together with Zagreb Rotunda (1937). Frei Otto developed the entire architectural vocabulary on pre-stressed cable nets and membranes and was in his early years in contact with Laffaille. With tensegrity systems a new step can be reached: using captors and actuators, it is now possible not only to have a pre- or self-stressed system, but also to modify the level and the distribution of this prescribed state of stress. This modification may help to adapt the system to its environment and to satisfy some criteria. This is known as *active control* and was illustrated in the chapter devoted to models.

7-34.2. Active control

Under the active impulsion of Mauro Pedretti and Professor Leopold Pflug, studies on active control have been undertaken at the "Ecole Polytechnique Fédérale de Lausanne" (Switzerland). The configuration of assembly of three cells (Figures 7.30 and 7.31) is identical to the one previously developed in Lugano by Passera & Pedretti. The node has been modified (Figure 7.32).

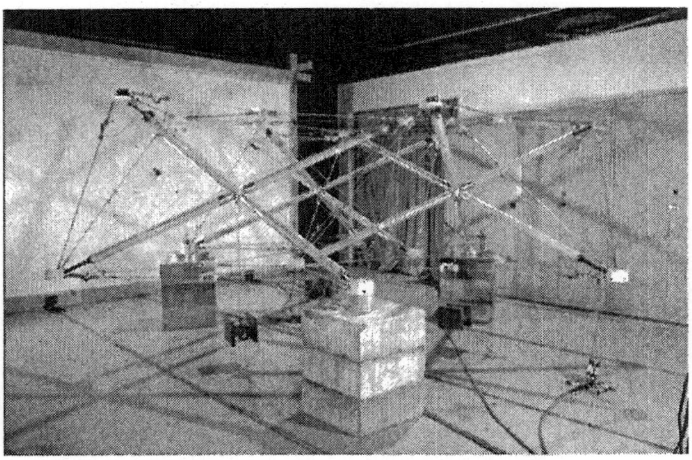

Figure 7.30 *Assembly of three cells – general view*

Figure 7.31 *Assembly of three cells*

Figure 7.32 *Central node*

Currently this project is being carried out by Etienne Fest with Ian Smith as supervisor.

7-34.3. Structures of the future

During the seminar in Tor Vergatà, Sergio Pellegrino described a radar project. In this project several characteristics of tensegrity systems are used: cable net membrane and struts are associated in a complex foldable system. This is an example of the possible applications of tensegrity in the field of the conquest of space.

The basic model is a hexagonal tensegrity prism. One of the layers is realised with a cable net, the other with a flexible membrane.

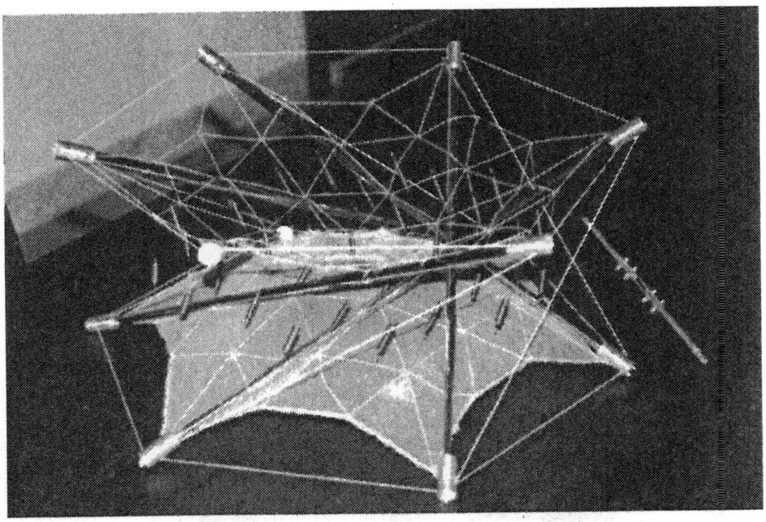

Figure 7.33 *Radar's model: view of the cable net*

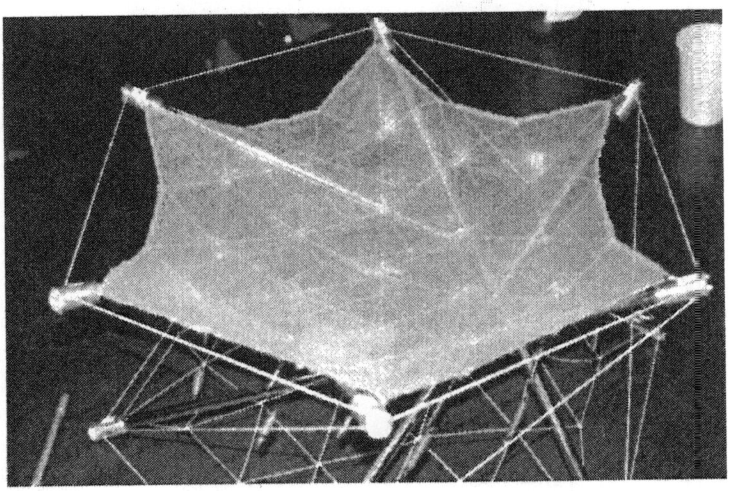

Figure 7.34 *Radar's model: view of the membrane*

The cable net and the membrane are linked with ties instrumented by small springs. The whole process of design has been published by G. Tibert [Ref 7-10].

7-4. Tensegrity as a structural principle

7-4.1. Introduction

As we come towards the end of this work on tensegrity, it appears that even if tensegrity systems have not yet been used in the field of architecture or building, the years to come will certainly be characterised by a major development in this regard. Furthermore, it is perhaps possible to extend the concept of tensegrity to other fields by analogy as Donald Ingber has already demonstrated in the field of biology.

I had the opportunity of participating in a seminar at the "College de Philosophie" in Paris. Tensegrity was the theme of the seminar. Donald Ingber and Patrizia d'Alessio spoke about tensegrity and biology and showed the components of the cytoskeleton (Figure 7.35) and also the progression of the cell (Figure 7.36) with adhesion description. They established the analogy with tensegrity and put emphasis on the form modification of cells from their beginning to their end.

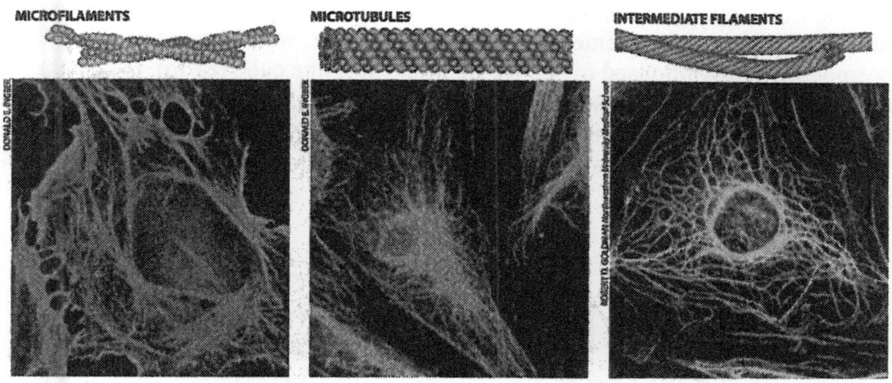

Figure 7.35 *Components of cytoskeleton*

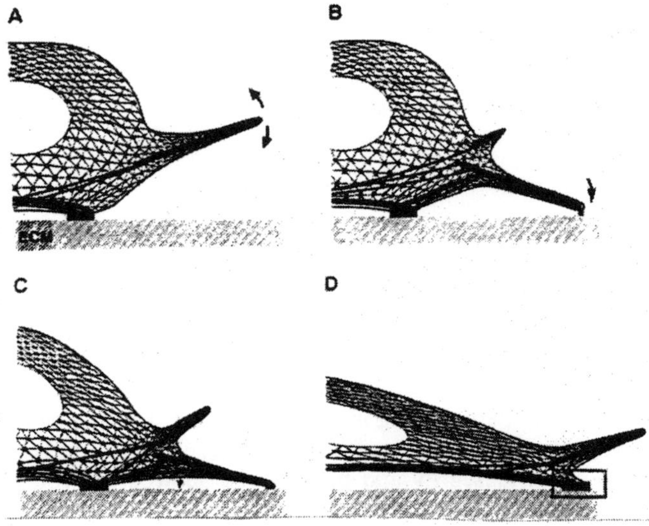

Figure 7.36 *Progression of the cell*

Simultaneously Luc Brisson, who is an eminent Hellenist, described the analogy set by Plato in Timée between the elements and the five regular polyhedra. It is well-known that the four elements that constitute the world according to Plato's cosmogony are symbolised by the tetrahedron (fire), the cube (earth), the octahedron (air) and the icosahedron (water). Plato claimed that four of them are included in the fifth one. I provided the geometric proof of this insertion in the so-called "ether" symbolised by the fifth platonician polyhedron, the dodecahedron [Ref 7-11] (see Figure 7.37).

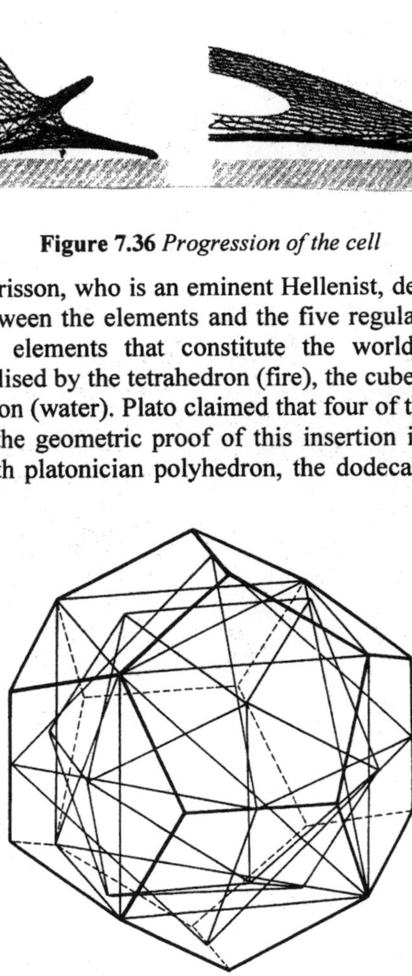

Figure 7.37 *Geometric insertion of the four elements in the "ether" symbolised by dodecahedron*

Closing the session Jean Dhombres enlightened the major role of geometric models. All these elements were very fruitful so far as tensegrity could be considered as a structural principle.

7-4.2. Coupling the visible and the invisible

A simple analogy cannot be considered superficially. The example of the arch is interesting. Let us consider a classical arch, and let us allow Marco Polo to say a few words:
Marco Polo describes a bridge stone by stone
But which is the stone which supports the bridge? Asks Kublai Khan.
The bridge is not supported by one or another stone, replies Marco Polo, but by the arch line that they constitute together.
Kublai Kahn remains silent. He is thinking. Then he adds:
Why do you speak me about stones? It is only the arch, which interests me.
Polo replies
Because, without stones, there is no arch.

Figure 7.38 *Arch at the entry of Olympia Stadium*

With these words Marco Polo was describing a coupling between forms and forces, between the visible and the invisible. But we have to keep in mind that this coupling between forms and forces also depends on materials and structure (this last word being taken in its relational meaning). One more time between the visible and the invisible. We have to take care of the limits of this process of analogy. It appears, for example, that it can be operated between the stone arch in Figure 7.38 and the slate arch in Figure 7.39.

Figure 7.39 *Slate arch (Andy Goldworthy)*

But, from a mechanical point of view, it is difficult to extend the analogy to a glass arch (Figure 7.40), since the freezing water acts to agglomerate the different pieces and transforms the system into a continuous system.

Figure 7.40 *Glass arch (Andy Goldworthy)*

7-4.3. Tensegrity as a structural principle

Even if caution is advised, it seems that a state of tensegrity can be considered as a structural principle which can be materialised or not. It corresponds to a specific field of forces, in a stable equilibrium, with a very specific distribution of the components which are always inside a continuum of tensions ("islands of compression in an ocean of tensions"). These forces are either compression or traction, and they can be associated with repulsion and attraction respectively. This enables the transposition of tensegrity to fields other than material ones.

7-5. Conclusion

As a conclusion I will take a quotation from Heraclites who claimed that the "**Universe is an harmony of tensions**" and described the Polemos principle as follows:

"*The supreme principle brings together conflicting opposites. It adapts the ones with the others in a balance, which is incessantly threatened by a dislocation risk. This law is valid in all fields at every level.*
*.... the same law of conflicting opposites adjustment, also called under the divine name of **Harmonia***".

References

Ref 7-1 Raducanu V., Architecture et système constructif: cas des systèmes de tenségrité, Thèse de doctorat, Université de Montpellier II, 2001.

Ref 7-2 Hanaor A., "Geometrically Rigid Double-layer Tensegrity Grids", *International Journal of Space Structures*, Vol. 9, N° 4, 1994, pp. 227–238.

Ref 7-3 Motro R., Joubert A., Systèmes de tenségrité à double nappe et double courbure: étude infographique et construction de maquette, Rapport Ministère de l'Equipement et du Logement, Direction de l'Architecture et de l'Urbanisme, B.R.A., 1990.

Ref 7-4 Motro R., Raducanu V., "Tensegrity systems", *International Journal of Space Structures*, to be published.

Ref 7-5 Pedretti M., "Smart tensegrity structures for the Swiss Expo 2001", Proceedings of SPIE, Vol. 3330, Washington, 1998, pp. 378–386.

Ref 7-6 Qirant J., Kébiche K., Kazi-Aoual M.N., "Etude des systèmes de tenségrité", Revue Française de Génie Civil, Vol. 4, 2000.

Ref 7-7 Raducanu V., Motro R., "Système à autoéquilibre stable pour élément de construction", demande de brevet français n° 01 04 822, déposée le 9 avril 2001 aux noms de C.N.R.S. – Ste Tissage et Enduction Serge Ferrari.

Ref 7-8 Kono Y., Choong K.K., Shimada T., Kunieda H., "An experimental investigation of a type of double layer tensegrity grids.", *Journal of IASS*, Vol. 41, n° 131, 2000.

Ref 7-9 Domer B., Fest E., "Active control of Tensegrity", Proceedings of Colloquium Lagrangianum Strutture tensegrity: analisi e progetti, Rome, 2001. To be published in "Revue Française de Génie Civil".

Ref 7-10 Tibert Gunnar, "Deployable Tensegrity Structures for Space Applications", Doctoral thesis, Stockholm, 2002, Royal Institute of Technology, Department of Mechanics.

Ref 7-11 Motro R., "Proportion and symbolism in Polyhedra", in Beyond the Cube: The Architecture of Space Frames and Polyhedra, edited by J. François Gabriel, ISBN 0-471-12261-0, Wiley & Sons Inc., 1997, pp. 281–299.

Appendices

A: Representation of nodes and actions

Reticulated system nodes are assimilated to material points that are represented by circles. These points are either free to be distributed in the space, either partially or totally fixed. Corresponding representations are illustrated in Figure A.1. First one is totally free, second free along X and Y directions, third totally fixed.

Figure A.1 *Convention for node representation*

Nodes are submitted to actions exerted by elements (Figure A.2) that are:

- either forces taking them for final end (centripetal), which are represented by black arrows; these forces are exerted by elements in a state of compression, which "push" the node.
- or forces taking the node for origin (centrifugal), which are represented by white arrows; these forces are exerted by elements in a state of tension, which "pull" the node.
- or external actions if any, which are represented by shadowed arrows.

The same convention is used for other solids.

Nodes are in equilibrium under the action of the totality of forces applied on them. Equilibrium of a system is equivalent to equilibrium of the totality of its nodes[1].

[1] This equilibrium can be studied algebraically or graphically with the help of dynamics of forces. The graphic method is well adapted to plan problems and to these that can be associated to them.

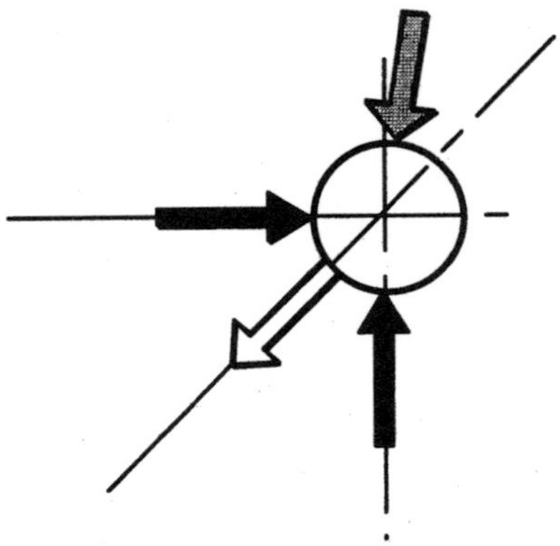

Figure A.2 *Actions exerted by elements on nodes*

B: Tension and compression

If one notes d_{ij} the effective distance between two nodes "i" and "j" in the system the element b of manufacture length l_{ij} will be[2]

"tensioned" if

E A-1 $l_{ij} < d_{ij}$

"compressed" if

E A-2 $l_{ij} > d_{ij}$

no stressed if

E A-3 $l_{ij} = d_{ij}$

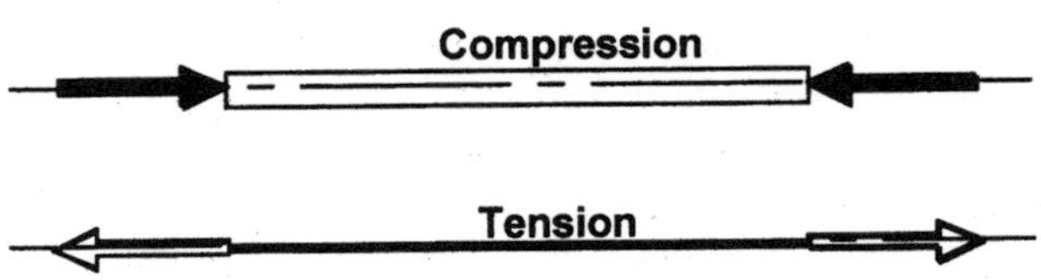

Figure A.3 *Tension-compression*

[2] These distances and these lengths are evaluated by assuming that nodes are points on which punctual extremities of elements are superposed. This hypothesis allows to forget real dimensions of the nodes.

C: Snelson's correspondence

Letter from Kenneth Snelson to René Motro, published in November 1990, in *International Journal of Space Structures*:

R. Motro
International Journal of Space Structures
Space Structures Research Centre
Department of Civil Engineering
University of Surrey, Guildford
Surrey GU2 5XH

Dear Mr. Motro:

I regret it has taken me so long to respond to your letter about the special issue of Space Structure dedicated to tensegrity.

As you probably know, I am not an engineer but an artist so I don't really feel qualified to write for an engineering journal. Nonetheless I know something about this particular form of structure from making so many sculptures over the years which use the principle which I prefer to call floating compression.

I have long been troubled that most people who have heard of "tensegrity" have been led to believe that the structure was a Bucky Fuller invention, which it was not. Of course, we are now in the year 1990 and not 1948 so all of this fades into the dim footnotes of history. There is a line somewhere in a theater piece which goes, "But that was long ago in another land -- and besides, the wench is dead."

Whenever an inventor defends his authorship the issue invariably turns out to be important only to the author himself, to others it is trivia. Maybe you're acquainted with the tale of Buckminster Fuller and me, but I'd like, somehow, to set the record straight, especially because Mr. Fuller, during his long and impressive career, was strong on publicity and, for his own purposes, successfully led the public to believe tensegrity was his discovery. He spoke and wrote about it in such a way as to confuse the issue even though he never, in so many words, claimed to have been its inventor. He talked about it publicly as "my tensegrity" as he also spoke of "my octet truss". But since he rarely accredited anyone else for anything, none of this is all that surprising. What Bucky did, however, was to coin the word tensegrity as he did octet truss and geodesic dome, dymaxion, etc., a powerful strategy for appropriating an idea. If it's his name, isn't it his idea?

As many new ideas do, the "tensegrity" discovery resulted in a way from play; in this case, play aimed at making mobile sculptures. A second-year art student at the University of Oregon in 1948, I took a summer off to attend a session in North Carolina at Black Mountain College because I had been excited by what I had read about the Bauhaus. The attraction at Black Mountain was the Bauhaus master himself, the painter Josef Albers who had taught at the German school and immigrated to the U.S. in 1933 to join the faculty of that tiny liberal arts college (fifty students that summer) in the Blue Mountains of North Carolina, fifteen miles from Asheville.

Buckminster Fuller, unknown to most of us in those early days, turned up two weeks into the session, a substitute for a professor of architecture who canceled a week before the summer began. Josef Albers asked me to assist the new faculty member in assembling his assortment of geometric models for his evening lecture to the college. There was no such thing as a tensegrity or discontinuous compression structure in his collection, only an early, great circle, version of his geodesic dome. Albers picked me to help because I had shown special ability in his three-dimensional design class.

During his lecture that evening Professor Fuller mesmerized us all with his ranging futurist ideas. As the summer quickly went by with most of the small school monitoring Fuller's classes I began to think I should try something three-dimensional rather than painting. Albers counseled me that I demonstrated talent for sculpture. But, more importantly, I had already become the first in a trail of students from colleges and universities who, over the years, were to become electrified "Fullerites". He had that cult-master's kind of charisma. I blush for it now, but it was true. We were young and looking for great issues and he claimed to encompass them all.

At the end of the summer session, I returned home to Pendleton, Oregon. In my Fullerian trance the descent into the real world was greatly confusing. I spent the autumn at home, making my parents miserable by moping and spending hours in the basement, building things; small mobile sculptures mostly, using thread, wire, clay, metal from tin cans, cardboard, etc. I had learned much about geometry from Fuller as well as art and design from the Bauhaus. While Albers' teachings were imparted as useable ideas in public-domain, Bucky's lessons were laden somehow with the sense that the ideas were proprietary -- "his" geometry. I believed, literally, because he claimed so, that before Buckminster Fuller came along, no human had ever noticed, for example, that to inscribe the diagonals of the square faces of a cube was to define two interlocking tetrahedra within. Students joked that, after all, hadn't Bucky invented the triangle? None of us knew, for example, of Alexander Graham Bell's early space frames, nor anything at all about crystallography.

In the autumn of 1948, as I said, I made numbers of small studies. Were they structures or sculptures? They incorporated the attitudes of both Fuller and Albers. The three small works which are of interest here were concerned both with balance of successive modular elements hinged one-to-another and stacked vertically as seen in photo #1; and, later, suspended one-to-the-next by means of thread-slings as shown in photograph #2. They were, of course, but amplifications of the familiar balancing toys seen often in novelty shops. My small discoveries in these two pieces were logical enough, though one could imagine that they might just as well lead to something other than to the first tensegrity structure; perhaps to variations on Calder mobiles.

It was the effort to make the pieces move which resulted in their spinal-column, modular, property. If I pushed on them lightly or blew on them, they swayed gently in a snake-like fashion. In photo #2 one can see module-to-module sling tension members replacing the wire hinges connecting the modules shown in photo #1. I thought of these threads as adding a note of mystery, causing the connections to be more or less invisible, at least as invisible as marionette strings; an Indian rope trick.

One step leading to the next, I saw that I could make the structure even more mysterious by tying off the movement altogether, replacing the clay weights with additional tension lines to stabilize the modules one to another, which I did, making "X", kite-like modules out of plywood. Thus, while forfeiting mobility, I managed to gain something even more exotic, solid elements fixed in space, one-to-another, held together only by tension members. I was quite amazed at what I had done. Photo #3

Still confused about my purposes and direction in school, I enrolled for engineering that winter ('48-'49) at Oregon State College. The classes depressed me even further. I hated it and did very poorly. I corresponded with Bucky and I told about my dilemma and also sent photos of the sequence of small sculptures. He must have understood from the letter how confused and depressed I was at school for he suggested I return for another Black Mountain Summer Session.

When we got together again in June I brought with me the plywood X-Piece (#3). When I showed him the sculpture, it was clear from his reaction that he hadn't understood it from the photos I had sent. He was quite struck with it, holding it in his hands, turning it over, studying it for a very long moment. He then asked if I might allow him to keep it. It hadn't been my intention to part with it, but I gave it to him, partly because I felt relieved that he wasn't angry that I had employed geometry (Buckminster Fuller's geometry) in making art. That original small sculpture disappeared from his apartment, so he told me at the end of the summer.

Next day he said he had given a lot of thought to my "X-column" structure and had determined that the configuration was wrong. Rather than the X-module for compression members, they should be shaped like the central angles of a tetrahedron, that is like spokes radiating from the gravitational center, to the vertices of a tetrahedron. Of course the irony was that I had already used that tetrahedral form in my moving sculpture #2, and rejected it in favor of the kite-like X modules because they permitted growth along all three axes, a true space-filling system, rather than only along a single linear axis. Those were not yet the years when students easily contradicted their elders, let alone their professors.

Next day I went into town and purchased metal telescoping curtain rods in order to build the "correct" structure for Bucky. I felt a little wistful but not at all suspicious of his motive as he had his picture taken, triumphantly holding the new structure I had built.

The rest of the story is one of numerous photographs and statements in print, grand claims in magazine articles and public presentations. In Time magazine he declared that, with "his" tensegrity, he could now span the Grand Canyon. He also described it as a structure which grows stronger the taller you build it -- whatever that may have meant.

The absorption process began early, even though Bucky penned the following in a letter to me dated December 22, 1949:

> **"In all my public lectures I tell of your original demonstration of discontinuous - pressure - (com-pressure) and continuous tension structural advantage; - in which right makes light in a prototype structure, the ready reproduction of which, properly incorporated in fundamental structures, may advance the spontaneous good will and understanding of mankind by many centuries. The event was one of those 'It happened' events, but demonstrates how the important events happen where the atmosphere is most favorable. If you had demonstrated this structure to an art audience it would not have rung the bell that it rang in me, who had been seeking this structure in Energetic Geometry. That you were excited by the latter, E.G., into spontaneous articulation of the solution, also demonstrates the importance of good faith of colleagues of this frontier. The name of <u>Ken Snelson</u> [his underline] will come to be known as a true pioneer of the realized good life and good will."**

Bucky's warm and uplifting letter arrived about six months after I first showed him my small sculpture. In that it was dated three days before Christmas, I suppose he was in a festive, generous, mood. A year later, January 1951 he published a picture of the structure in Architectural Forum magazine and, surprisingly, I was not mentioned. When I posed the question some years later why he accredited me, as he said, in his public lectures and never in print, he replied, **"Ken, old man, you can afford to remain anonymous for a while."**

Finally, in 1959 I learned that Fuller was to have a show at the Museum of Modern Art in New York and included in it was to be a thirty-feet high tensegrity "mast". Calling it a mast seemed especially obtuse, but he regarded himself as a man of the sea. With some persistence and with the lucky aid of Bucky's assistant I was able to get word to Arthur Drexler, curator at the Museum, about my part in tensegrity. This forced Bucky's hand. At last, my credit for tensegrity found its way into the public record.

One of the ironies of this not-too-unusual tale in the history of teacher-student relationships, is that by Bucky's transposing my original "X" module into the central-angles-of-the-tetrahedron shape to rationalize calling it his own, he managed successfully to put under wraps my original form, the highly adaptable X form. He could not have lived with himself with the blatant theft of my original system, of course, and besides, he had denounced it as the "wrong" form. As a result, none of the many students in schools where he lectured ever got to see it. In those years, any number of students labored to constructed their own "masts", but all were built using the tetrahedral form. That moment of recognition at the Museum of Modern Art in November 1959, transitory as it was, was quite fortifying and enabled me to once again pick up my absorbing interest in this kind of structure with the feeling that now I was free and on my own. Especially I picked up where I had left off with the neglected X-module which was left unnoticed for an entire decade. I no longer felt anonymous.

As I said earlier, this is but a footnote to a storm in a teapot. I have continued to make sculptures which now stand in public sites in many places. Sorry there are none in England or France. The ghost of Bucky Fuller continues to muddy the water in regard to "tensegrity". I tell myself often that, since I know where the ideas came from, that ought to be enough.

As I see it, this type of structure, at least in its purest form is not likely to prove highly efficient or utilitarian. As the engineer Mario Salvadori put it to me many years ago, "The moment you tell me that the compression members reside interior of the tension system, I can tell you I can build a better beam than you can." He was speaking metaphorically about this type of structure in general, of course. Over the years I've seen numbers of fanciful plans proposed by architects which have yet to convince me there is any advantage to using tensegrity over other methods of design. Usually the philosophy is akin to turning an antique coffee-grinder into the base for a lamp: it's there, so why not find a way to put it to some use. No, I see the richness of the floating compression principle to lie in the way I've used it from the beginning, for no other purpose than to unveil the exquisite beauty of structure itself. Consciously or unconsciously we respond to the many aspects of order in nature. For me, these studies in forces are a rich source for an art which celebrates the aesthetic of structure, of physical forces at work; force-diagrams in three-dimensional space, as I describe them.

Whether or not you are able to use this narrative about the beginnings of tensegrity, I wish you the very best with your special issue on the subject.

Sincerely, Kenneth Snelson

Photo #1 (Snelson, 1948)

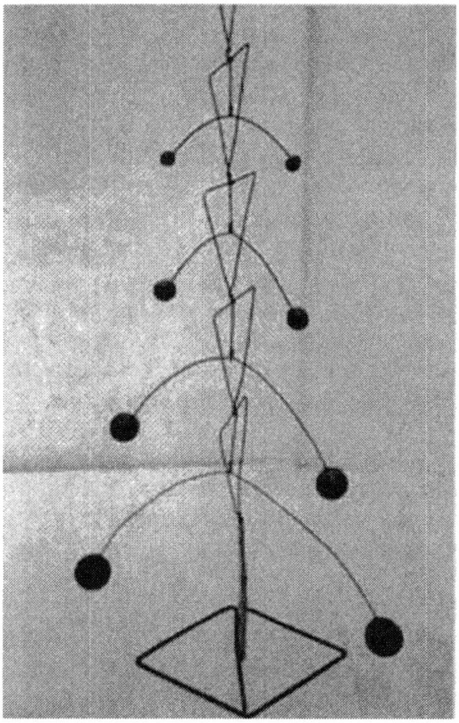

Photo #2 (Snelson, 1948)

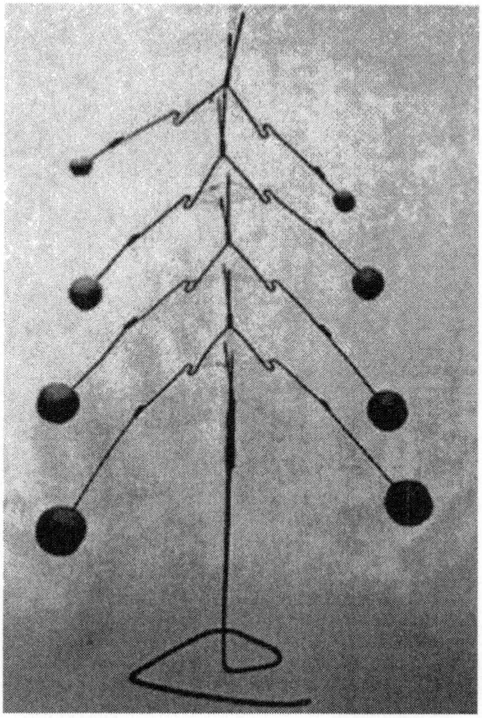

Photo #3 (Snelson, 1948)

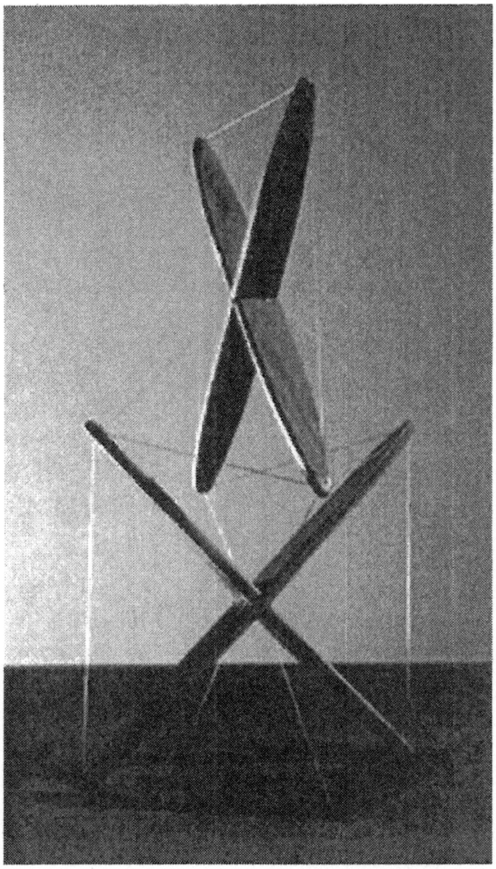

Bibliography

1959 1. Emmerich D.G., "Charpentes Perles" ("Pearl Frameworks"), Institut National de la Propriété Industrielle (Registration n° 59423), 26 May 1959.

1965 2. Snelson K., "Continuous tension, discontinuous compression structures" U.S. Patent No 3,169,611, 16 February 1965.

1967 3. Emmerich D.G., "Réseaux", in Space Structures: A study of methods and developments in three-dimensional construction, Proceedings of the International Conference on Space Structures, Guildford 1966, edited by R.M. Davies, Blackwell Scientific Publications, 1967, pp. 1059–1072.

4. Emmerich D.G., Cours de Géométrie Constructive – Morphologie, École Nationale Supérieure des Beaux Arts, Paris, Centre de diffusion de La Grande Masse, Paris, 1967.

1969 5. Graybill, Introduction to matrices with applications in statistics, Wrels Worth Publishing Company, Inc., Belmont, California, 1969.

1970 6. Boulet P., "Mât Chinois", Exercices de Géométrie Constructive, École Nationale Supérieure des Beaux Arts Paris, Presses du Centre de Recherches et d'Urbanisme, 1970, p. 93.

7. Chassagnoux A., "Mât Autotendant", Exercices de Géométrie Constructive, École Nationale Supérieure des Beaux Arts Paris, Presses du Centre de Recherches et d'Urbanisme, 1970, p. 211.

8. Emmerich D.G., Exercices de Géométrie Constructive, École Nationale Supérieure des Beaux Arts Paris, Presses du Centre de Recherches et d'Urbanisme, 1970.

1971 9. Linkwitz K., Sheck H.J., "Einige Bemerkungen von vorgespannten Seilnetzkonstruktionen", Ingenieur Archiv 40, Springer Verlag, 1971, pp. 145–158.

1972 10. Yao J.T.P., "Concept of structural control", *Journal of the Structural Division, ASCE*, Vol. 98, N° ST7, Proc. Paper 9048, July 1972, pp. 1569–1570.

1973 11. Fuller R.B., Marks R., The Dymaxion World of Buckminster Fuller, Anchor Books, 1973.

12. Minke G., "Tensegrity System", Revue Zodiac, N° 19, 1973.

13. Motro R., Topologie des structures discrètes. Incidence sur leur comportement mécanique. Autotendant icosaédrique. Note de

recherche interne 74, Laboratoire de Génie Civil, Université des Sciences, Montpellier, 1973.

1974 14. Linkwitz K., Sheck H.J., "The force density method for formfinding and computation of networks", *Computer Methods in Applied Mechanics and Engineering*, 3, 1974, pp. 115–134.

1975 15. Bathe K.J., Ramm E., Wilson E., "Finite Element Formulations for Large Deformation Dynamic Analysis", *International Journal for Numerical Methods in Engineering*, Vol. 9, 1975, pp. 353–385.

16. Fuller R.B., Synergetics explorations in the geometry of thinking, Collier Macmillan Publishers, London, 1975.

17. Tardiveau J., Siestrunck, R., "Efforts et déformations dans les assemblages en treillis critiques et surcritiques en élasticité linéaire", Note présentée par M.M. Roy à l'Académie des Sciences, T. 280, N° 3, 20 January 1975.

1976 18. Kenner H., Geodesic maths and how to use it, University of California Press, Berkeley, 1976.

19. Pugh A., An introduction to tensegrity, University of California Press, Berkeley, 1976.

20. Pugh A., Polyhedra, a visual approach, University of California Press, Berkeley, 1976.

1978 21. Abdel-Rohman M., Leipholz H., "Active Control of Flexible Structures Dynamic Analysis", *Journal of the Structural Division, ASCE*, Vol. 104, N° ST8, Proc. Paper 13964, August 1978, pp. 1251–1266.

1981 22. Roth B., Whiteley W., "Tensegrity Frameworks", Transactions of the American Mathematical Society, Vol. 256, N° 2, 1981, pp. 419–446.

1982 23. Abdel-Rohman M., Leipholz H., "Active Control of Large Structures", *Journal of the Engineering Mechanics Division, ASCE*, Vol. 108, N° EM5, October 1982.

24. Bathe K.J., Finite Element Procedures, Prentice Hall, Englewood Cliffs, New Jersey, 1982.

1983 25. Abdel-Rohman M., Leipholz H., "Active Control of Tall Buildings", *Journal of Structural Engineering*, Vol. 109, N° 3, March 1983.

26. Motro R., Formes et Forces dans les Systèmes Constructifs. Cas des Systèmes Réticulés Spatiaux Autocontraints, Thèse d'Etat, Université Montpellier II, June 1983.

1984 27. Motro R., "Review of the development of geodesic domes", in Analysis, Design and Construction of Braced Domes, edited by Z.S. Makowski, Granada Publishing Ltd., 1984, pp. 387–411.

28. Nooshin H., Formex configuration processing in structural engineering, Elsevier, ISBN 0-85334-315-2, 1984.

1985 29. Fuller R.B., Inventions: The Patented works of R.B. Fuller, St. Martin's Press, 1985.

1986 30. Edmondson A., "Geodesic Reports: The Deresonated Tensegrity

Dome", *Synergetica Journal of Synergetics*, Vol. 1, N° 4, November 1986.

31. Motro R., Najari S., Jouanna P., "Tensegrity Systems from Design to Realisation.", Proceedings of The First International Conference on Lightweight Structures in Architecture, Sydney, Australia, edited by V. Sedlak, 1986, pp. 690–697.

32. Pellegrino S., Calladine, "Matrix analysis of statically and kinematically indeterminate frameworks", in *International Journal of Solids and Structures*, Vol. 22, 1986, pp. 409–428.

1987

33. Belkacem S., Recherche de forme par relaxation dynamique de systèmes réticulés spatiaux autocontraints, Thèse de Docteur Ingénieur, Université Paul Sabatier, Toulouse, 1987.

34. Maculet R., Etude et calcul de structures autotendantes, Diplôme d'Architecture, École d'Architecture de Paris La Villette, 1987.

35. Motro R., "Tensegrity Systems for Double Layer Space Structures", Proceedings of the International Conference on the Design and Construction of Non Conventional Structures, edited by B.H.V. Topping, London, 1987, pp. 43–51.

36. Reinhorn A.M., Manolis G.D., Wen C.Y., "Active Control of Inelastic Structures", *Journal of Engineering Mechanics*, Vol. 113, N° 4, 1987.

1988

37. Cha J.Z., Pitarresi J.M., Soong T.T., "Optimal Design Procedures for Active Structures", *Journal of Engineering Mechanics*, Vol. 114, N° 12, 1988.

38. Chu R., "Tensegrity", *Journal of Synergetics*,Vol. 2, N° 1, June 1988.

39. Emmerich D.G., Structures tendues et autotendantes – Monographies de Géométrie Constructive, Éditions de l'Ecole d'Architecture de Paris La Villette, 1988.

40. Rampon A., Formes et structures, Dossier personnel d'études, Ecole d'architecture Languedoc Roussillon, 1988.

41. Soong, T.T., "Active Structural Control in Civil Engineering", *Engineering Structures*, Vol. 10, 1988.

1989

42. Chung L.L., Lin R.C., Soong T.T., "Experimental Study of Active Control for MDOF Seismic Structures", *Journal of Engineering Mechanics*, Vol. 115, N° 8, 1989.

1990

43. Eco U., Le Pendule de Foucault, ©Gruppo Editoriale Fabbri, Bompiani, Songzono, Etas S.pA., Milan, 1988, © Editions Grasset et Fasquelle, 1990.

44. Motro R., Joubert A., Systèmes de tenségrité à double nappe et double courbure: étude infographique et construction de maquette, Rapport Ministère de l'Equipement et du Logement, Direction de l'Architecture et de l'Urbanisme, B.R.A., 1990.

45. Motro R., "Tensegrity Systems and Geodesic Domes", Special Issue of the *International Journal of Space Structures*, "Geodesic Space Structures", Vol. 5, N° 3, 1990, pp. 343–354.

46. Yang J.N., Akbarpour, Askar G., "Effect of Time Delay on Control of Seismic-Excited Buildings", *Journal of Structural Engineering*, Vol. 116, N° 10, 1990.

1991

47. Hangai Y., "Analytical procedure for stabilising paths and stability of kinematically indeterminate frameworks", in Proceedings of International Association for Spatial and Shell Structures Symposium 1991, Copenhagen, 1991.
48. Yang J.N., Danielians A., Liu S.C., "A Seismic Hybrid Control Systems for Building Structures", *Journal of Engineering Mechanics*, vol. 117, N° 4, 1991.

1992

49. Burdon R., Laxton John, "Foldable structures", UK Patent application GB 2 256 444 A, 09.12.1992.
50. Chang C.C., Yang H.T.Y., "Instantaneous Optimal Control of Building Frames", *Journal of Engineering Mechanics*, vol. 120, N° 4, 1992.
51. Furuya H., "Concept of deployable tensegrity structures in space application", in *International Journal of Space Structures*, Vol. 7 N° 2, 1992, pp. 143–151.
52. Hanaor A., "Aspects of Design of Double-layer Tensegrity Domes" *International Journal of Space Structures*, Vol. 7, N° 2, 1992, pp. 101–114.
53. Lodder C., The Transition to Constructivism. The Great Utopia. The Russian and Soviet Avant-Garde, 1915–1932, exhibition, Guggenheim Museum, 1992.
54. Motro R., "Tensegrity Systems: State of Art", *International Journal of Space Structures* (Special Issue on Tensegrity Systems), R. Motro Guest Editor, Vol. 7, N° 2, 1992.
55. Yang J.N., Li Z., Liu S.C., Yang J.N., "Control of Hysteretic System Using Velocity and Acceleration Feedbacks", *Journal of Engineering Mechanics*, Vol. 118, N° 11, 1992.
56. Yang J.N., Li, Z., and Liu, S.C., "Stable Controllers for Instantaneous Optimal Control", *Journal of Engineering Mechanics*, Vol. 118, N° 8, 1992.

1993

57. Calatrava S., Sobre la plegabilidad de entramados, In Arquitectura Transformable, Publicacion de la Escola Tecnica Superior de Arquitectura de Sevilla, 1993, pp. 33–93.
58. Hanaor A., "Double-layer tensegrity grids as deployable structures", in *International Journal of Space Structures*, Vol. 8, N° 1&2, 1993, pp. 135–145.
59. Kwan A.S.K, Pellegrino S., "Design and performance of the octahedral deployable space antenna (Odesa)", in *International Journal of Space Structures*, Vol. 9, N° 3, 1993, pp. 163–173.
60. Kwan A.S.K, You Z., Pellegrino S., "Active and passive cable elements in deployable/retractable masts", in *International Journal of Space Structures*, Vol. 8, N° 1/2, 1993, pp. 29–40.
61. Rofooei F.R., Tadjbakhsh I.G., "Optimal Control of Structures with

Acceleration, Velocity, and Displacement Feedback", *Journal of Engineering Mechanics*, Vol. 119, N° 10, 1993.

1994 62. Bouderbala M., Systèmes à barres et câbles pliables, dépliables, Diplôme d'Etudes Approfondies, Université Montpellier II, 1994.

63. Hanaor A., "Geometrically rigid double-layer tensegrity grids", *International Journal of Space Structures*, Vol. 9, N° 4, 1994, pp. 227–238.

1995 64. Contemporary Developments in Design Sciences, exhibition, Cathedral of St John the Divine, New York, USA, November 1995.

65. Liu Y., Motro R., "Shape analysis and internal forces in unstable structures", Proceedings of the IASS International Symposium 1995 "Spatial Structures: Heritage, Present and Future", edited by G.C. Giuliani, SGE pub., Padovia, Vol. 2, 1995, pp. 819–826.

1996 66. "Morphology", *International Journal of Space Structures* (Special Issue on Morphology), H. Lalvani Guest Editor, Vol. 11, N° 1 & 2, 1996.

67. Motro R., Bouderbala M., "Mobile Tensegrity Systems", in Proceedings of MARAS 96 "Mobile and Rapidly Assembled Structures II", 1996, pp. 103–113.

68. Motro R., "Structural Morphology of Tensegrity Systems", *International Journal of Space Structures* (Special Issue on Morphology), H. Lalvani Guest Editor, Vol. 11, N° 1 & 2, 1996, pp. 233–240.

1997 69. Ingber D.E., "Tensegrity: The Architectural Basis of Cellular Mechanotransduction", *Annu. Rev. Physiol.*, 1997, pp. 575–599.

70. Motro R., "Proportion and Symbolism in Polyhedra", in Beyond the Cube: The Architecture of Space Frames and Polyhedra, edited by J. François Gabriel, ISBN 0-471-12261-0, Wiley & Sons Inc., 1997, pp. 281–299.

71. Nooshin H., Disney P.L., Champion O.C., "Computer-Aided Processing of Polyhedric Configurations", in Beyond the Cube: The Architecture of Space Frames and Polyhedra, edited by J. François Gabriel, ISBN 0-471-12261-0, Wiley & Sons Inc., 1997, pp. 281–299.

72. Raducanu V., Assemblages: conception d'un nœud industriel pour les grilles de tenségrité, Diplôme d'Etudes Approfondies. Université Montpellier II, 1997.

73. Vassart N., Recherche de forme et stabilité des systèmes réticulés autocontraints, Thèse de doctorat, Université des Sciences et Techniques du Languedoc, Montpellier, 1997.

1998 74. Bouderbala M., Systèmes spatiaux pliables/dépliables: cas des systèmes de tensegrité, Thèse de doctorat, l'Université de Montpellier II, 1998.

75. Djouadi S., Le contrôle des structures et les systèmes de tenségrité, Thèse de doctorat, Université des Sciences et Techniques du Languedoc, Montpellier, 1998.

76. Djouadi. S., Motro R., Pons J.C., Crosnier B., "Active control of tensegrity systems", *Journal of Aerospace Engineering*, Vol. 11, N° 2, ASCE, ISSN 0893; Paper N° 16880, 1998, pp. 37–44.
77. Ingber D.E., "The Architecture of Life", *Scientific American*, January 1998, pp. 30–39.
78. Kebiche K., Etude en non linéarités géométriques et homogénéisation des systèmes réticulés spatiaux autocontraints, Thèse de doctorat, Université Montpellier II, 1998.
79. Le Saux C., Modélisation numérique du pliage des systèmes de tenségrité avec prise en compte des contacts, Mémoire de DEA, Université de Montpellier II, 1998.
80. Pedretti M., "Smart Tensegrity Structures For The Swiss Expo 2001", Proceedings of LSA98 "Lightweight structures in architecture engineering and construction", edited by Richard Hough & Robert Melchers, ISBN 0 9586065 0 1, Vol. 2, 1998, pp. 684–691.
81. Pedretti M., "Smart tensegrity structures for the Swiss Expo 2001", Proceedings of SPIE, Vol. 3330, Washington, 1998, pp. 378–386.
82. Wang B.B., Yan Yun Li, "Definition of tensegrity systems. Can dispute be settled?" Proceedings of LSA98 "Lightweight structures in architecture engineering and construction", edited by Richard Hough & Robert Melchers, ISBN 0 9586065 0 1, Vol. 2, 1998, pp. 713–719.
83. Williamson D., Skelton R.E, "A general class of tensegrity systems: Equilibrium analysis", in Engineering Mechanics for the 21st Century, SCE Conference, La Jolla, 1998.

1999

84. Kebiche K., Kazi-Aoual M.N., Motro R., "Geometrical non-linear analysis of tensegrity systems", *Journal of Engineering Structures*, 21, 1999, pp. 684–876.
85. Hangai Y., Wu M., "Analytical method of structural behaviours of a hybrid structure consisting of cables and rigid structures", *Engineering Structures*, 21, 1999, pp. 726–736.
86. Sultan C., Modeling, design, and control of tensegrity structures with applications. Ph.D. Dissertation, Purdue University, School of Aeronautics and Astronautics, West Lafayette, 1999.
87. Vassart N., Motro R., "Multiparametered Formfinding Method. Application to Tensegrity Systems", *International Journal of Space Structures* (Special Issue on Formfinding of Tension Systems), Vol. 14, N° 2, 1999, pp. 131–146.
88. Volokh K. Yu., "Nonlinear analysis of underconstrained structures", *International Journal of Solids and Structures*, 36, 1999, pp. 2175–2187.
89. Wemyss G.A., Support Structure. UK Patent Application GB 2 331 456 A. 26.05.1999.

2000

90. Ben Kahla N., Kébiche K., Non linear elastoplastic analysis of tensegrity systems, *Engineering Structures*, 23, 2000, pp. 1552–

1566.

91. Kangwai R.D., Guest S.D., "Symmetry-adapted equilibrium matrices", *International Journal of Solids and Structures*, 37, 2000, pp. 1525–1548.
92. Kono Y., Choong K.K., Shimada T., Kunieda H., "An experimental investigation of a type of double layer tensegrity grids", *Journal of IASS*, Vol. 41, N° 131, 2000.
93. Qirant J., Kébiche K., Kazi-Aoual M.N., "Etude des systèmes de tenségrité", *Revue Française de Génie Civil*, Vol. 4, 2000.
94. Quirant J., Kazi Aoual N., Motro R., "Tensegrity Systems: Selfstress States and Sensitivity", in Proceedings of IASS-IACM 2000. Fourth International Colloquium on Computation of Shell and Spatial Structure, Chania Crete, 2000, pp. 170–171.
95. Quirant J., Systèmes de Tenségrité et autocontrainte: qualification, sensibilité et incidence sur le comportement. Thèse de doctorat, Université des Sciences et Techniques du Languedoc, Montpellier, 2000.
96. Sultan C., Corless M., Skelton R.E., "Tensegrity flight simulator", *Journal of Guidance, Control, and Dynamics*, 23(6), 2000, p. 1055.
97. Vassart N., Laporte R., Motro R., "Determination of mechanisms's order for kinematically and statically indeterminate systems", *International Journal of Solids and Structures*, Vol. 37, 2000, pp. 3807–3839.
98. Volokh K. Yu., Vilnay O., Belsky M., "Tensegrity architecture explains linear stiffening and predicts softening of living cells", *Journal of Biomechanics*, 33, 2000, pp. 1543–1549.
99. Volokh K. Yu., Vilnay O., "Plane frames as semi-underconstrained systems", *International Journal of Mechanical Sciences*, 42, 2000, pp. 1119–1134.

2001

100. de Jager B., Skelton R.E., "Optimizing stiffness properties of tensegrity structures", in Proceedings of International Mechanical Engineering Congress and Exposition, Vol. 3330, New York, 2001.
101. Domer B., Fest E., "Active Control of Tensegrity", Proceedings of Colloquium Lagrangianum Strutture tensegrity: analisi e progetti, Rome, 2001. To be published in *Revue Française de Génie Civil.*
102. Foucher O., Polyèdres et Tenségrité, Mémoire de DEA, Université Montpellier II, 2001.
103. Micheletti A., "On the kinematics of tensegrity towers", Colloque Lagrange, Rome, 2001
104. Murakami H., Nishimura Y, "Static and dynamic characterisation of regular truncated icosahedral and dodecahedral tensegrity modules", to appear in *International Journal of Solids and Structures*, 2001.
105. Raducanu V., Architecture et système constructif: cas des systèmes de tenségrité, Thèse de doctorat, Université de Montpellier II, 2001.
106. Raducanu V., Motro R., "Système à autoéquilibre stable pour

élément de construction", demande de brevet français n° 01 04 822, déposée le 9 avril 2001 aux noms de C.N.R.S. – Ste Tissage et Enduction Serge Ferrari.

107. Skelton R.E.,. Pinaud J.P, Mingori D.L., "Dynamics of the shell class of tensegrity structures", *Journal of the Franklin Institute*, 338/2–3., 2001, pp. 255–320.

108. Smaili A., Fuller Snelson Emmerich, Mémoire de DEA, Université Montpellier II, 2001.

109. Sultan C., Corless M., Skelton R.E. "The prestressability problem of tensegrity structures, some analytical solutions". To appear in *International Journal of Solids and Structures*, 2001.

2002

110. Motro R. and Raducanu V., "Tensegrity systems", *International Journal of Space Structures*, in press.

111. Tibert Gunnar, "Deployable Tensegrity Structures for Space Applications", Doctoral thesis, Stockholm, 2002, Royal Institute of Technology, Department of Mechanics.

Index

FSC
www.fsc.org
MIX
Papier | Fördert gute Waldnutzung
FSC® C083411

Kolibri 360 GmbH
Ferdinand-Jühlke-Straße 7
99095 Erfurt, Deutschland
produktsicherheit@kolibri360.de